SpringerBriefs in Optimization

SpringerBriefs in Optimization showcases algorithmic and theoretical techniques, case studies, and applications within the broad-based field of optimization. Manuscripts related to the ever-growing applications of optimization in applied mathematics, engineering, medicine, economics, and other applied sciences are encouraged.

More information about this series at http://www.springer.com/series/8918

Francesco Archetti · Antonio Candelieri

Bayesian Optimization and Data Science

Springer

Francesco Archetti
Department of Computer Science
Systems and Communications
University of Milano-Bicocca
Milan, Italy

Antonio Candelieri
Department of Computer Science
Systems and Communications
University of Milano-Bicocca
Milan, Italy

ISSN 2190-8354 ISSN 2191-575X (electronic)
SpringerBriefs in Optimization
ISBN 978-3-030-24493-4 ISBN 978-3-030-24494-1 (eBook)
https://doi.org/10.1007/978-3-030-24494-1

This Springer imprint is published by the registered company Springer Nature Switzerland AG
The registered company address is: Gewerbestrasse 11, 6330 Cham, Switzerland

Preface

“*Data culture is decision culture*” it’s hard to disagree with this statement in Mckinsey quarterly (2018, 3); the point is further stressed by the “takeaway”: “*Don’t approach data analysis as a cool science experiment or an exercise in amassing data for data sake. The fundamental objective in collecting, analyzing and deploying data is to make better decisions*”.

This is an old objective, first put forward by the operations research community: linear programming was in use, as an operational decision tool, already in Dantzig[1] and around the same time the dynamic programming framework was proposed.

Optimization methods have since moved well beyond the traditional area of engineering design and operations management and have become ubiquitous in all branches of science, business and government.

It is also increasingly clear that in a data-driven environment, most of the times we are called to take decisions in conditions of partial information and uncertainty: decision-making, as data becomes available, bearing in mind the impact of our decisions in the different futures which might unfold, raises new challenges and requires a different mindset.

Data-driven optimization also arises when the objective function and/or constraints are analytically unknown and need to be learned during the search for the optimal solution: evaluating the objective function in these cases might be very expensive, like building a prototype of a system, drilling a borehole looking for ore, treating a patient or simulating, ab initio, the pharmacological activity of a new chemical compound. Also more common applications, like designing, training and validating a machine learning application on a large dataset and optimizing its configuration and hyperparameters fall into a simulation–optimization or black-box framework and can be computationally challenging.

As in any learning process, a central issue is the so-called exploration *vs* exploitation dilemma, which will be recurrent in this monograph: exploration means devoting resources to learn about the structure of the problem, in particular

[1]Dantzig, G.B.: Alternate algorithm for the revised simplex method: Using a product form for the inverse. Rand. (1953)

possible solutions, while exploitation devotes resources to improve on solutions already identified in the previous phase.

An algorithm suitable for such applications should, above all, be **sample efficient,** because the cost of observations—which amount to function evaluations—is the dominating cost and have **global** properties.

These requirements have important consequences:

1. To be sample efficient it must query, at a relatively low cost, the design space to find the new point with maximal informative value.
2. The search for the new point must strike an effective balance between the needs of exploration and exploitation.

We shall see how the Bayesian optimization (BO) framework allows to formulate a mathematically principled way to satisfy these requirements.

We shall refer to this behaviour of searching for highly informative new data as "active learning".

"Learning enabled" optimization requires a model flexible enough to adjust its parameters to new data, monitoring on line its performances, and be discriminating, sifting through data streams in order to select from them which to harvest and analyse on the basis of their informational utility.

Learning has been studied in many communities notably statistics and cognitive sciences.

Among the statisticians, the Bayesians believe that all kinds of learning underpin Bayes' theorem and developed a branch of statistical thinking whose application, due to computational difficulties, has been originally limited to relatively simple problems. Things began to change due to the recent availability of computing power and new computational methods which enabled the application of Bayesian learning in large scale meaningful problems and its spreading to many domains including machine learning (the big success of naïve Bayesian classifier in spam classification) and optimization.

The Bayesian algorithmic optimizer updates its beliefs about the shape of the objective function, the location the value of its global optimum and even the number of local optima according to the evidence gathered thru function evaluations. To do this, we need a model which can sum up our a priori beliefs (before fresh data arrives) and update them as new data arrives, as mandated by the Bayes' theorem:

$$P(beliefs \mid data) = P(beliefs) \times P(data \mid beliefs)/P(data)$$

The models learned of the objective function are called also response surface or surrogate model: they are only approximations of different "fidelity" of the objective function, allow to account for the uncertainty in its estimation and can drive in a principled way the search towards the optimum.

Bayesian optimization is so far the best computational strategy at the cross-roads between learning and optimization, exploring the environment or the space of decision variables to gather efficiently new relevant information and to exploit it towards optimal solutions.

All this seems natural and easy, yet to bring BO into action did require a number of advances in software engineering, computational sciences and statistical learning which have been critical for its successful deployment in many application domains. The modelling and computational difficulties as well as the methods and tools developed to overcome them are the subject of this book.

Who discovered Bayesian Optimization? Of course, Thomas Bayes[2] in the founding paper of Bayesian statistics, but it is fair to remind what a contemporary of Bayes', the Scottish philosopher David Hume wrote in his 1748 essay "An enquiry concerning human understanding": "*a wise man proportions his beliefs to the evidence*" a statement fitting squarely in the Bayesian framework.

Closer to our time, the idea was put forward in Kushner[3] and Mockus[4] although, not without reason, it could be partly traced back to a paper 1951 of Krige[5] and its subsequent development largely in the engineering community, under the name of kriging, for simulation-based optimization.

Another class of algorithms, originated from the work of Strongin[6] and called information statistical approach, was further developed, alongside with space covering methods, and has consolidated the foundations of Lipschitz Global Optimization.

BO quickly became one of the main strategies for global optimization, a domain which received first systematic attention in the two volumes[7,8] and whose meaning was originally the optimization of non-convex /multi-extremal functions and later grew to include simulation-based and black-box optimization.

An important point was the paper of Jones[9] which started to consolidate BO as a design tool in the engineering community and to spread it to the machine learning research community[10] where it has become the "de facto" standard for automatic

---

[2]Bayes, T. 1764. "An Essay Toward Solving a Problem in the Doctrine of Chances", *Philosophical Transactions of the Royal Society of London* **53**, 370–418.

[3]Kushner, Harold J. "A new method of locating the maximum point of an arbitrary multipeak curve in the presence of noise". Journal of Basic Engineering 86, no. 1 (1964): 97–106.

[4]Močkus, J. (1975). On Bayesian methods for seeking the extremum. In Optimization Techniques IFIP Technical Conference (pp. 400–404). Springer, Berlin, Heidelberg.

[5]Krige, D. G., 1951, A statistical approach to some basic mine valuation problems on the Witwatersrand: J. Chem. Metal. Min. Soc. South Africa, v. 52, pp. 119–139.

[6]Strongin, R.G.: Multiextremal minimization of measurements with interference. Eng. Cybern. 16, 105–115 (1969).

[7]Dixon, L. C. W., Szegö, G. P.: Towards Global Optimisation: Proceedings of a Workshop at the University of Cagliari, Italy, October 1974. (1975).

[8]Dixon, L. C. W., Szegò, G. P.: Towards global optimisation. North-Holland Amsterdam (1978).

[9]Jones, Donald R., Matthias Schonlau, and William J. Welch. "Efficient global optimization of expensive black-box functions". Journal of Global optimization 13, no. 4 (1998): 455–492.

[10]Bergstra, J. S., Bardenet, R., Bengio, Y., & Kégl, B. (2011). Algorithms for hyper-parameter optimization. In Advances in neural information processing systems (pp. 2546–2554).

algorithm configuration and hyperparameter optimization as we shall explain in Chap. 1.

BO, as we remarked before, is based on a surrogate model: true as it is that "all models are wrong, some are useful", the Gaussian processes offer the model that turned out more useful as a probabilistic surrogate in BO. Gaussian processes are a powerful nonparametric formalism for implementing both regression and classification algorithms which also offer a reliable uncertainty estimate. Two early papers on Bayesian Optimization are also Archetti[11], with an analysis of unidimensional Bayesian Optimization, and Archetti and Betrò[12], for a general convergence analysis of sequential optimization of measurable random functions.

One way to interpret a Gaussian process (GP) regression model is to think of it as defining a distribution over functions, and with inference taking place directly in the space of functions. A Gaussian process is a collection of random variables, any finite number of which have a joint Gaussian distribution. A Gaussian process is completely specified by its mean function $\mu(x) = \mathbb{E}[f(x)]$ and covariance function $k(x, x') = \mathbb{E}[(f(x) - \mu(x))(f(x') - \mu(x'))]$ also named *kernel* which amounts to impose a geometry on our decision space (alike for support vector machines). Given a prior belief, for instance on the basis of a random sampling, the Gaussian process provides an analytical expression for the update of mean and variance (posterior) as new function evaluations arrive which in turn allows to build the analytic expression of an "acquisition " function which suggests the new observation points and modulates the trade-off between exploration and exploitation.

Who should read this book?

This monograph can be useful for many kinds of readers but was basically written having in mind three main audiences:

1. graduate and doctoral students, mostly in computer science, artificial intelligence and optimization, who want to become knowledgeable in paradigmatic learning enabled optimization method which has become the "de facto standard" for hyperparameter optimization and algorithm configuration in the machine learning community.
2. software engineers, who do not have a training in machine learning, but are using industrial platforms, and want to understand the inside workings of BO as a service for automated machine learning and black-box optimization.
3. researchers, from both academia and industry who look for a quick understanding of whether and how their specific activity can benefit from BO as a tool

[11]Archetti, F., Betrò, B.: A probabilistic algorithm for global optimization. Calc. *16*, 335–343 (1979). https://doi.org/10.1007/BF02575933

[12]Archetti, F., Betro, B.: Stochastic models and optimization. Boll. Della Unione Mat. Ital. *5* (17-A), 295 (1980)

for the design of experiments and the optimization of computationally expensive simulation models.

The contents of this monograph are organized as follows.

Chapter 1 outlines the basic structure of BO with reference to the problem of automated machine learning and the design of predictive analytics pipeline.

Chapter 2 shows how BO relates to global optimization, in particular Lipschitz-based methods and Random Search. Section 2.4, in particular, shows how BO could be regarded, along with approximate dynamic programming, an early instance of learning enabled optimization, tightly knitted with key artificial intelligence frameworks like a multi-armed bandit and reinforcement learning.

Chapter 3 presents the key components of a Gaussian process analysing several widely used kernels and some computational issues. Anyway one should not see GP as one-size-fit-all recipe: other models have been recently suggested including Bayesian neural networks and deep neural networks which are presented in Sects. 3. 2 and 3.3.

Chapter 4 is dedicated to the different acquisition functions both the traditional and the new ones, based on information theory which have been proposed to cope with theoretical developments and emerging application fields.

Chapter 5 is considering “exotic” instances of BO with a specific reference to constrained BO, where a novel approach has been proposed by the authors of this book to cope with unknown constraints and partially defined objective functions.

Chapter 6 is a survey of the software resources available for BO, analysed with respect to the following features: surrogate model, acquisition functions, language, licence and last update.

Finally, Chap. 7 is a survey of applications in many industrial sectors: BO has been in the last decade a burgeoning field largely due to the work of the machine learning community, but its features have led to its adoption by different communities in several innovative applications: from the challenge of AlphaGO[13] to “machine learning for optimizing cookie recipe”[14], through the operational optimization of pump scheduling in water distribution systems, the design of drugs and new materials, robotics and industrial automation, to name just a few, Bayesian optimization has become a ubiquitous presence any time a machine learning application has to be designed according to sample efficiency and performance criterion.

Different reading paths can be obtained linking different chapters: Chaps. 1 and 6 are enough to set up a “naïve”, problem agnostic BO solution for a given machine learning application or a black-box optimization problem.

[13]Chen, Yutian, Aja Huang, Ziyu Wang, Ioannis Antonoglou, Julian Schrittwieser, David Silver, and Nando de Freitas. “Bayesian optimization in AlphaGo”. arXiv preprint arXiv:1812.06855 (2018).

[14]Solnik, Benjamin, Daniel Golovin, Greg Kochanski, John Elliot Karro, Subhodeep Moitra, and D. Sculley. “Bayesian Optimization for a Better Dessert”. (2017).

Adding Chaps. 3 and 4 gives the modelling and computational tools to tailor the BO solution that fits one's problem and software resources: these four chapters are the "core" of the book.

Outside the core, Chap. 2 gives a conceptual framework to link BO with other learning enabled optimization approaches. One should read it before Chap. 5 to grasp the implications of the Lipschitz condition in "safe" BO.

This book is designed as a self-contained entry into the BO field. The coverage of the topics, but the basic ones, is quite sketchy. The amount of theoretical basis is minimal, just enough to keep it self-contained. Some more advanced contents are exposed dealing with Thompson sampling in Chap. 3.

Citations are limited by the space constraints: so, the coverage of literature is extensive but by no means complete. Albeit the purpose of this book is to provide a working methodology, some references to incremental improvements and hints at a historical perspective have been regarded by the authors as possibly helpful and have been given in Chap. 2.

Milan, Italy

Francesco Archetti
Antonio Candelieri

Acknowledgements The authors want to acknowledge the contribution of Riccardo Perego, Stanislav Fedorov, Bruno G. Galuzzi and Ilaria Giordani who read the book, provided useful suggestions to several chapters and performed valuable computational experiments. Another acknowledgement to Yaroslav Sergeyev of University of Calabria and Anatoly Zhjigliavsky of Cardiff University: discussions with them, besides providing valuable suggestions, were encouraging and motivating.

Although this work is "brief" it exacted anyway its toll in terms of time taken from our families over evenings and weekends. Francesco A. is thankful to Stefania for her tenderness and unfaltering support; Antonio C. to his prior Valentina and their major update Samuel.

Contents

Chapter 1
Automated Machine Learning and Bayesian Optimization

1.1 Automated Machine Learning

1.1.1 Motivation

Automated machine learning (AutoML) is introduced in this chapter to illustrate model selection and hyperparameter tuning and the specific features of the resulting optimization problems.

These issues are important because an algorithm that scores well on one learning task can score poorly on another, as summarized by the "no free lunch" (NFL) theorem (Wolpert and Macready 1995): even the same algorithm, with different hyperparameter values, can show very different performances (Wolpert 2002). NFL theorems have since grown into a significant research domain: their impact on fields like optimization and supervised learning is analysed in a recent survey paper (Adam et al. 2019), in which specific results are given for early stopping and cross-validation rules to conduct the training phase.

While this result emphasizes the role of the ML experts in the design of reliable applications, substantial research activity has been focusing on the possibility to exploit the potential of ML by making it easy to be used off the shelf also by non-experts, taking, in some sense, the "human out of the loop" (Shahriari 2016). This requires automating the configuration of the algorithm and the tuning of their hyperparameters.

The time-honoured approach to solve the problem of hyperparameter optimization, namely grid search, is hardly feasible for more than 2/3 hyperparameters and totally unsuitable in view of the growing complexity of ML models as deep networks which have many hyperparameters and may take hours or days to train the model.

Model selection and hyperparameter optimization are very important applications of Bayesian optimization (BO) (Frazier 2018) and offer a good motivation of the relevance assumed by BO in the ML community (Hutter et al. 2019). Section 1.2 is devoted to illustrating the basic workflow of the sequential model-based optimization (SMBO) and specifically of BO introducing its basic components: the surrogate

F. Archetti and A. Candelieri, *Bayesian Optimization and Data Science*, SpringerBriefs in Optimization, https://doi.org/10.1007/978-3-030-24494-1_1

model and the acquisition function. Section 1.3 is devoted to a specific application showing the design and the implementation of a complex ML application of predictive analytics.

AutoML was formulated in Kotthoff et al. (2017) as "automatically and simultaneously choosing a learning algorithm and setting its hyperparameters to optimize its empirical performance on a given dataset". The more general definition of this problem is also called combined algorithm selection and hyperparameter (CASH) optimization.

This approach was originally built on WEKA and developed into Auto-WEKA (Thornton et al. 2013) which uses BO for an optimal selection among the components of WEKA for a given dataset and problem addressed (regression or classification). More recently, a new and robust AutoML system auto-sklearn, based on scikit-learn, has been proposed in Feurer et al. (2015).

This system uses several supervised learning algorithms, data preprocessing and feature processing methods, leading to an overall number of up 110 machine learning algorithms which can be compared according to their generalization capability that is to make accurate "prediction" (regression or classification) on new data.

The evaluation of a ML algorithm means evaluating any performance index or metric (e.g. accuracy or root-mean-squared error in classification and regression tasks, respectively) which we shall indicate as loss function.

This evaluation is usually done by a k-fold cross-validation procedure. This procedure consists in randomly dividing the set of data into k groups, or *folds*, of approximately equal size. With an index i iterating from 1 to k, the ith fold is used as a validation or testing set while the ML algorithm is trained on the training set consisting of the remaining $k-1$ folds. This means that each algorithm must be trained and validated k times, making this procedure time consuming and significantly expensive, especially in the case of large databases or for those algorithms whose training is expensive. Usually, a limited "budget" to train and test different configurations of an ML algorithm is available, bounding the available computational resources such as CPU, wall clock time, memory usage.

The successful application of ML in a broad range of domains and the shortage of human experts has recently led to an increase in the demand of AutoML solutions. Indeed, a growing number of commercial enterprises are addressing this demand: Microsoft's Azure Machine Learning, Amazon Machine Learning and Google's platforms (Prediction API, Vizier, Hypertune and the more recent AutoML Tables) offer "machine learning" ecosystems with their own AutoML tools. The need of sample efficient AutoML solutions cannot be overestimated also in view of the growing awareness of the heavy carbon footprint of large scale machine learning models, in particular deep learning. A recent paper (Strubell et al. 2019) argues that the cost of the hyperparameter tuning and the required experimentation (Sect. 1.1.3) can reach a tag of CO_2 emissions in the range of tens of thousands Libs which can reach hundreds of thousands if the neural architecture search (Sects. 1.1.2 and 1.1.4) is also considered. The unprecedented scale of this challenge requires not only better algorithms in the Bayesian framework but revolutionary solutions like quantum computing. Bayesian Optimization, (Zhu et al. 2018), has turned out a good solution for data -driven quantum circuit training, using the software tool OPTaas, of Mindfoundry which is analysed in Chap. 6.

1.1.2 Model Selection

A learning algorithm A maps training data points $x_1, \ldots, x_n$ to their corresponding "targets" $y_1, \ldots, y_n$, where y_i is continuous in the case of regression or a "label" (categorical values) in the case of classification. All the pairs (x_i, y_i) are organized into a dataset $\mathcal{D} = \{(x_i, y_i)\}_{1:n}$.

The model selection problem is formulated as follows:

$$A^* \in \underset{A \in \mathcal{A}}{\operatorname{argmin}} \frac{1}{k} \sum_{i=1}^{k} \mathcal{L}\left(A, \mathcal{D}_{\text{train}}^{(i)}, \mathcal{D}_{\text{valid}}^{(i)}\right)$$

where $\mathcal{L}\left(A, \mathcal{D}_{\text{train}}^{(i)}, \mathcal{D}_{\text{valid}}^{(i)}\right)$ is the loss achieved by A when trained on $\mathcal{D}_{\text{train}}^{(i)}$ and evaluated on $\mathcal{D}_{\text{valid}}^{(i)}$. The set $\mathcal{A}$ contains all the available ML algorithms, $\mathcal{A} = \left\{A^{(1)}, \ldots, A^{(h)}\right\}$. We use k-fold cross-validation which splits the training data into k equal-sized validation folds, $\mathcal{D}_{\text{valid}}^{(1)}, \ldots, \mathcal{D}_{\text{valid}}^{(k)}$, and associated training sets $\mathcal{D}_{\text{train}}^{(1)}, \ldots, \mathcal{D}_{\text{train}}^{(k)}$, where $\mathcal{D}_{\text{train}}^{(i)} = \mathcal{D} \backslash \mathcal{D}_{\text{valid}}^{(i)}$.

The following figure (Fig. 1.1) displays a schematic representation of a model selection problem.

Before going ahead, it is important to clarify the difference between "training" and "validating" a ML model, as well as the difference between "parameters" and "hyperparameters" of a ML algorithm.

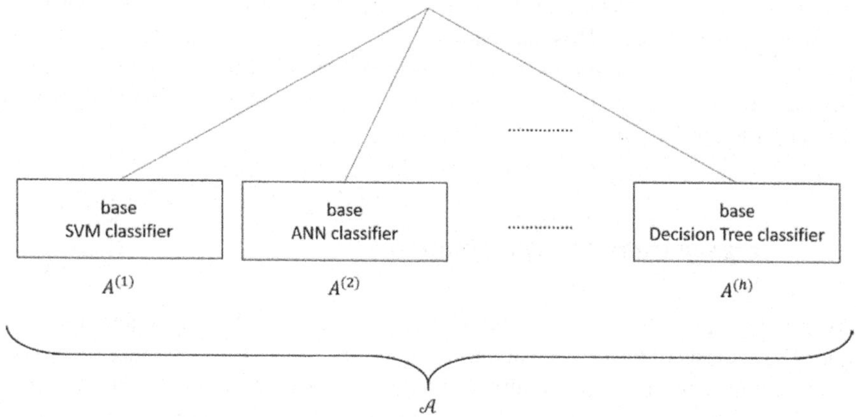

Fig. 1.1 An example of model selection. A set of "base" algorithms are available in the set A, where "base" refers to the adoption of default values for the algorithm's hyperparameters

The algorithm A is characterized by:

- a vector θ of "model parameters", whose value is learned directly from data, during the "training" phase
- a vector of "hyperparameters" $\gamma \in \Gamma$ that change the way the algorithm "learns" the values for θ. Hyperparameters can be set up manually or optimized on the "validation" phase.

For instance, the weights on the connections of an artificial neural network are parameters to be learned, while a number of hidden layers, a number of neurons in each hidden layer, activation function and learning rate are hyperparameters.

The goal of *training* is to estimate the value $\hat{\theta}$ minimizing a given loss function $\mathcal{L}_{\text{train}}$ (e.g. classification error or root-mean-squared error for classification and regression problems, respectively) on training data $\mathcal{D}$. The resulting ML model can be then validated on a validation dataset (or fold, in the case of k-fold cross-validation), measuring the corresponding loss function $\mathcal{L}_{\text{valid}}$. During validation, the estimate $\hat{\theta}$ does not change for each fold. The hyperparameters γ are not estimated during the train, and they must be set up before training. Therefore, $\hat{\theta}$ as well as $\mathcal{L}_{\text{train}}$ and $\mathcal{L}_{\text{valid}}$ depend on the value of γ.

Many ML algorithms, such as in artificial neural networks or support vector machines, use an analytical form for $\mathcal{L}_{\text{train}}$, so that $\hat{\theta}$ can be estimated by gradient-based methods. Other ML algorithms have no parameters (e.g. instance-based algorithms such as k-nearest neighbours). On the contrary, $\mathcal{L}_{\text{valid}}$ is black box and its optimization, depending on the hyperparameters γ, requires derivative-free GO approaches, such as BO.

When k-fold cross-validation is concerned, the common way to compute the overall $\mathcal{L}_{\text{valid}}$ is to average the values obtained on the k different folds, as in a randomized experiment.

The loss function computed on each fold $\mathcal{L}^i_{\text{valid}}$ with $i = 1, \ldots, k$, can be considered as an observation from a sample of k elements, whose size can be controlled by some statistical tests. This approach could result in an early stopping when the hyperparameter configuration is unlikely to yield a good result (Florea et al. 2018). The same principle is used in hyperband (Li et al. 2016), halving sequentially the set of configurations "deserving" more sampling.

1.1.3 Hyperparameter Optimization

Given n hyperparameters $\gamma_1, \ldots, \gamma_n$ with domains $\Gamma_1, \ldots, \Gamma_n$, the hyperparameter space Γ is a subset of the product of these domains $\Gamma_1 \times \cdots \times \Gamma_n$. Γ is a subset because certain settings of one hyperparameter render other hyperparameters inactive. For example, the parameters determining the specifics of the third layer of an artificial neural network (ANN) are not relevant if the network depth is set to one or two.

Likewise, the hyperparameters of a support vector machine's (SVM) polynomial kernel are not relevant if we use a different kernel instead. More formally, we say that a hyperparameter γ_i is conditional on another hyperparameter γ_j; that is γ_i is

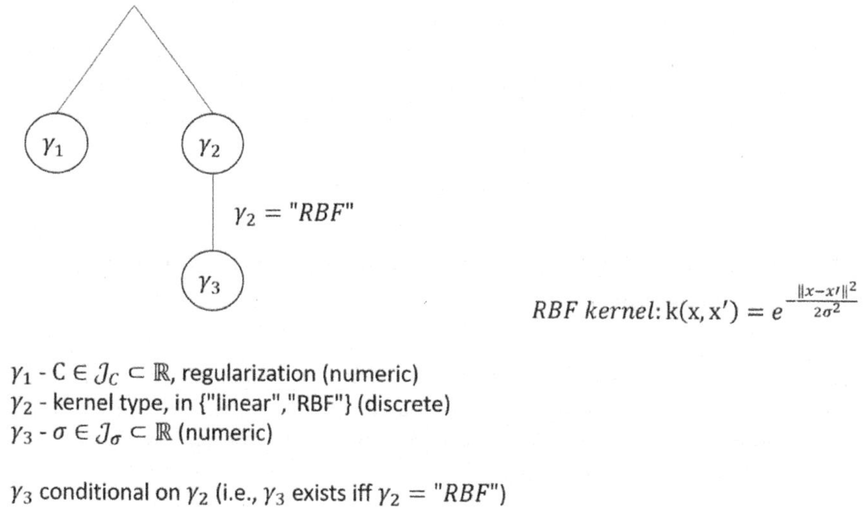

Fig. 1.2 An example of hyperparameter optimization for a support vector machine (SVM) classifier with linear or radial basis function (RBF) kernel. $\mathcal{J}_C$ and $\mathcal{J}_\sigma$ are the ranges for values of the regularization hyperparameter C and the RBF kernel's hyperparameter σ

active if and only if hyperparameter γ_j takes values from a given set $V_i(j) \subset \Gamma_j$; in this case, we call γ_j a parent of γ_i. Conditional hyperparameters generate a tree-structured space or, in some cases, a directed acyclic graph (DAG). Given such a structured space Γ, the (hierarchical) hyperparameter optimization problem can be written as:

$$\gamma^* \in \operatorname{argmin}_{\gamma \in \Gamma} \frac{1}{k} \sum_{i=1}^{k} \mathcal{L}\left(A_\gamma, \mathcal{D}_{\text{train}}^{(i)}, \mathcal{D}_{\text{valid}}^{(i)}\right) \tag{1.1}$$

Figures 1.2 and 1.3 propose two examples of hyperparameter optimization for an SVM and an ANN, respectively.

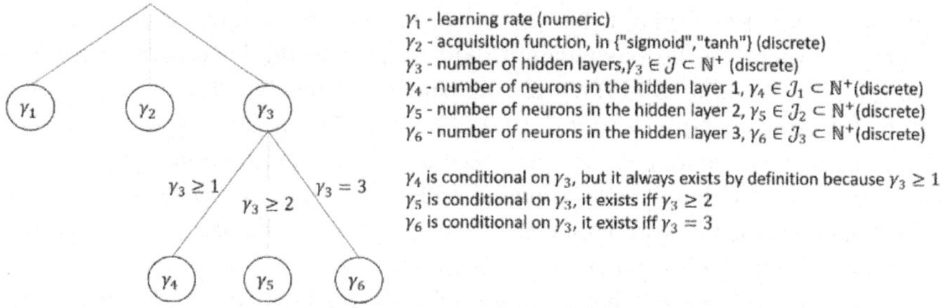

Fig. 1.3 An example of hyperparameter optimization for an artificial neural network (ANN) classifier with at maximum three hidden layers. $\mathcal{J}_1$, $\mathcal{J}_2$ and $\mathcal{J}_3$ are the ranges for number of neurons in the hidden layer 1, 2 and 3, respectively

The problem of hyperparameter optimization of a given learning algorithm is quite similar to model selection. There are still some key differences in that hyperparameters are often continuous, that hyperparameter spaces are often high dimensional, like in deep neural networks, and that we can exploit correlation structure between hyperparameter settings $\boldsymbol{\gamma}, \boldsymbol{\gamma}' \in \Gamma$.

1.1.4 Combined Algorithm Selection and Hyperparameter Optimization

Given a set of algorithms $\mathcal{A} = \{A^{(1)}, \ldots, A^{(h)}\}$ with associated hyperparameter spaces, $\Gamma^{(1)}, \ldots, \Gamma^{(h)}$, we define the combined algorithm selection and hyperparameter optimization problem (combined algorithm selection and hyperparameter optimization, CASH) as computing:

$$A^*_{\gamma^*} \in \operatorname{argmin}_{A^{(j)} \in \mathcal{A}, \gamma \in \Gamma^{(j)}} \frac{1}{k} \sum_{i=1}^{k} \mathcal{L}\left(A^{(j)}_{\gamma}, \mathcal{D}^{(i)}_{\text{train}}, \mathcal{D}^{(i)}_{\text{valid}}\right) \tag{1.2}$$

We note that this problem can be reformulated as a single combined hierarchical hyperparameter optimization problem with hyperparameter space $\Gamma = \Gamma^{(1)} \cup \ldots \cup \Gamma^{(h)} \cup \{\gamma_r\}$, where γ_r is a new root-level hyperparameter that selects between algorithms $A^{(1)}, \ldots, A^{(h)}$. The root-level hyperparameter of each subspace $\Gamma^{(i)}$ is made conditional on γ_r. In principle, the problem can be tackled in various ways: SMBO is a versatile stochastic optimization framework that can work with both categorical and continuous hyperparameters and that can exploit the hierarchical structure stemming from the conditional parameters (Kotthof et al. 2017) (Fig. 1.4).

1.1.5 Why Hyperparameter Optimization Is Important?

Figures 1.5 and 1.6 display a loss function, namely mean squared error (MSE), computed on ten fold cross-validation, with respect to the hyperparameters of a support vector machine (SVM) regression, namely the regularization $C \in [0, 1000]$ and the hyperparameter $\sigma \in [0.002, 0.1]$ of the radial basis function (RBF) kernel. Two different benchmark datasets have been used: CPU dataset (https://archive.ics.uci.edu/ml/datasets/Computer+Hardware) and Yacht dataset (http://archive.ics.uci.edu/ml/datasets/yacht+hydrodynamics). One can see how the performance metrics (i.e. MSE) is impacted by hyperparameter value.

Another critical feature is that the value of the loss function, for each hyperparameter configuration, is the outcome of the randomized process of k-fold cross-validation. An often-overlooked point is to measure the uncertainty of the prediction error esti-

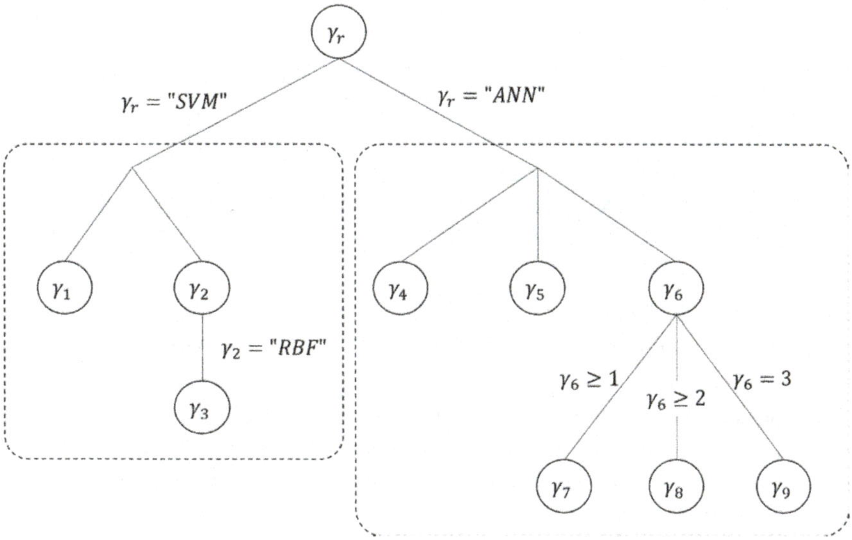

Fig. 1.4 An example of CASH considering an SVM and an ANN classifier. Description of the hyperparameters of each algorithm follows from Figs. 1.2 and 1.3, while γ_r is the further "root" CASH hyperparameter introduced to model the choice between the types of machine learning algorithm

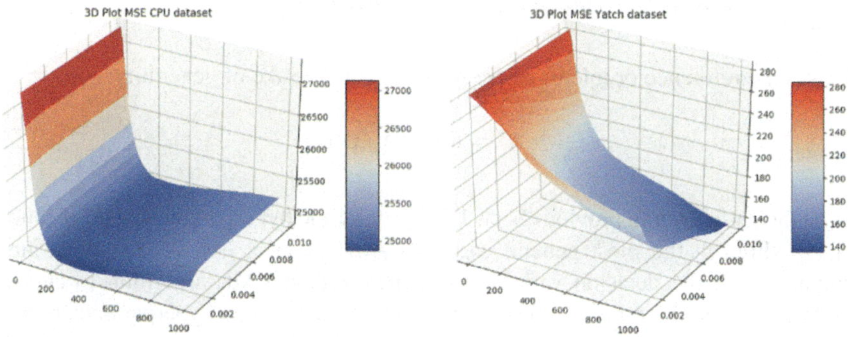

Fig. 1.5 Error on ten fold cross-validation with respect to two hyperparameters of an SVM with RBF kernel on the CPU dataset (left) and Yacht dataset (right)

mators which is important because the accuracy of model selection is limited by the variance of model estimates. Cross-validation provides an unbiased estimate of the prediction error on the training set, but the estimation of the variance is still crucial. Seminal contributions to this problem are given in Nadeau and Bengio (2000) and Bengio and Grandvalet (2004).

In a recent paper (Jiang and Wang 2017), a uniform normalized variance is proposed which not only measures model accuracy but relates it to the number of folds.

Let us have a look at the simplest case in which we keep fixed the value of C of an SVM and can choose among different kernels, represented by the options A to E.

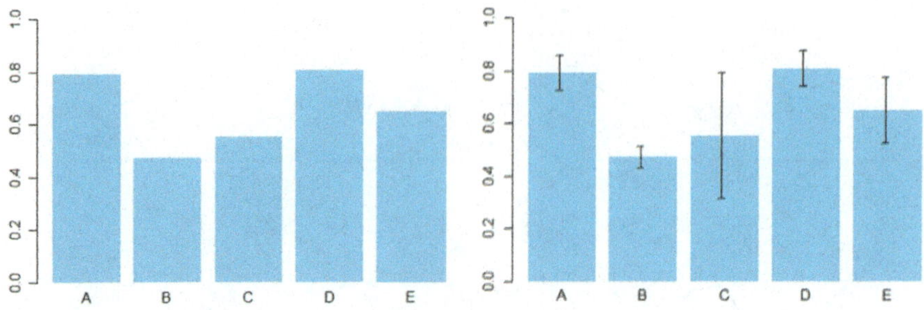

Fig. 1.6 MSE computed over k-fold cross-validation for an SVM classifier with five different kernels (options A to E). On the left, the mean of MSE is reported, and on the right also variance is depicted

If we consider the mean of MSE over the k-folds, we obtain the chart on the left, but we ignore the variance, which is instead depicted in the figure on the right. The variance makes the choice not so obvious. Indeed, there is a nonzero probability that option C could turn out a better choice than E.

1.2 The Basic Structure of Bayesian Optimization

1.2.1 Sequential Model-Based Optimization

This is the reference problem, where we use x for the sake of generality:

$$\mathbf{x}^* = \arg\min_{\mathbf{x}\in X} f(\mathbf{x})$$

In the cases seen in previous Sect. 1.1 $\mathbf{x} = \boldsymbol{\gamma}$ for the hyperparameter optimization (1.1), while, in the most general case of CASH (1.2), $\mathbf{x}$ is a vector whose first component is the decision variable associated with the ML algorithm to choose in the set $\mathcal{A}$ and the remaining components are the associated hyperparameters $\boldsymbol{\gamma}$.

When we are dealing with this optimization problem, we are faced with two kinds of uncertainty:

1. Measurement uncertainty called "noise". We can not observe $f(x_i)$ but rather a noisy value $y_i = f(x_i) + \varepsilon$, where $\varepsilon$is the observation noise, assumed to be $\varepsilon \sim \mathcal{N}(0, \lambda^2)$ (Fig. 1.7).

Many papers are dealing with noise (Frazier 2018), but most of results in BO can be obtained without explicitly taking measurement noise into account. The "workhorse" GP model can handle it without disrupting the basic algorithm structure.

2. Even removing the noise, as it is often assumed, we still have a problem of "structural uncertainty". For instance, if we have three noise-free evaluations of $f(x)$, $D_{1:3} = \{(x_i, y_i)\}_{i=1,\dots,3}$, we still have infinite number of functions with different minima and minimizers, compatibly with $D_{1:3}$, as depicted in Fig. 1.8.

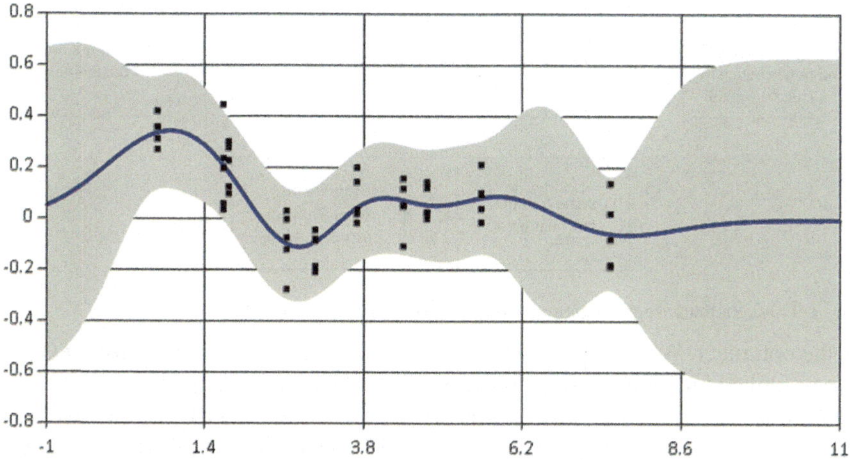

Fig. 1.7 An example of GP model with noisy evaluation: for the same point, x_i is possible to observe different observations y_i

Fig. 1.8 Different functions compatible with the function observations $D_{1:3}$

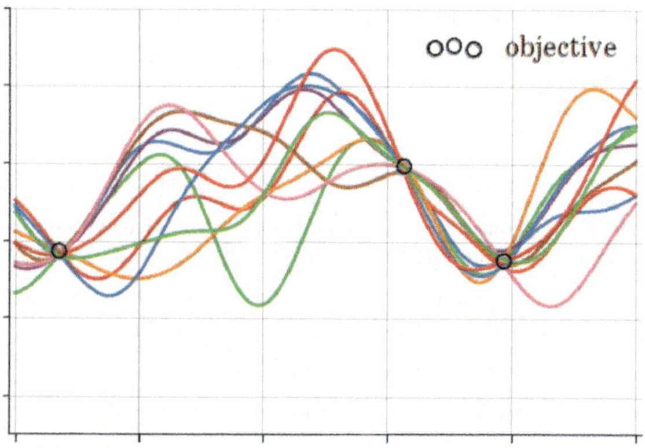

In many cases, X is assumed a hyper–rectangle (box-bounded or essentially unconstrained optimization). In this chapter, we present the basic structure of BO in which the design variables, the hyperparameters, are assumed to be continuous.

We remark that BO can handle more complex design space: variables can be continuous, integer, categorical and conditional. This is in general the case of automatic algorithm configuration (AAC), hyperparameter optimization in machine learning and CASH.

Nonlinear constraints can be considered including the case in which they are the results of a simulation model and are partially/completely unknown, as will be discussed in Chap. 5.

We do not assume the existence in $f(x)$ of structural properties like linearity/convexity. We also assume that first and second derivatives are not available

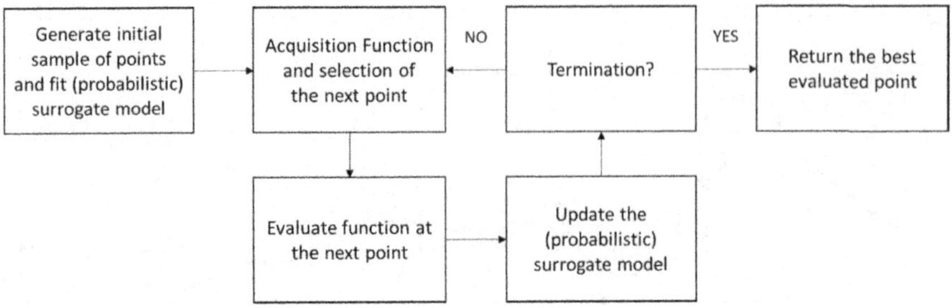

Fig. 1.9 A schematic representation of the Bayesian optimization process

to the optimizer. Where derivative information can be observed along with $f(x)$, it can be embedded in the learning process, as we shall see in Chap. 3.

In situations as the above, an approach is to create a surrogate model of the objective function and build on it a sequential model-based optimizer (SMBO) evaluating $f(x)$ at the points $x_1, x_2, \ldots, x_n$ updating, on the basis of new values, the surrogate and, as shown in Figs. 1.10 and 1.11, suggesting the next evaluation point. This suggestion is performed by an acquisition function, which balances exploitation and exploration (as better detailed in Sect. 1.2.3) and represets the "utility" to select a point. The SMBO process is summarized in the schema and steps (Fig. 1.9).

Steps of the sequential model-based optimization process
STEP 1—Generating initial sample/design—by Random Search (Chap. 2) or other sampling methods—and fitting the first (probabilistic) surrogate model (Sect. 1.2.2 and Chap. 3)
STEP 2—Identifying, depending on the acquisition function, the new promising point(s), evaluating the function (Sect. 1.2.3 and Chap. 4)
STEP 3—Evaluating the objective function
STEP 4—Update the (probabilistic) surrogate model (Chap. 3)
STEP 5—Checking for termination criteria: if at least one is active, go to STEP 6, else go to STEP 2
STEP 6—Return the current best solution (usually the "best seen")

1.2.2 Surrogate Model

In this book we focus on probabilistic surrogate model, we must anyway remark that deterministic surrogate models are also widely studied and applied in academia and industry (Cozad et al. 2014). Probabilistic surrogate models are able to offer an estimate of the uncertainty which is used to balance the trade-off between exploration and exploitation. The Gaussian process is most widely adopted model. The idea behind GP is that the values of the objective function correspond to realizations of a multivariate Gaussian process that for each x returns a mean and a variance, as reported in Figs. 1.10 and 1.11.

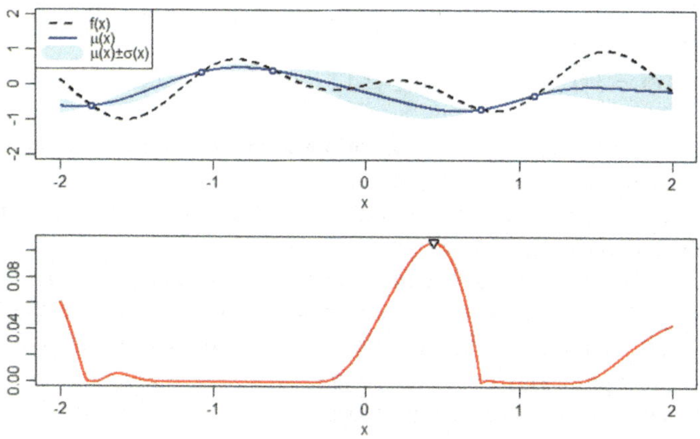

Fig. 1.10 A GP conditioned to five observations (top) and corresponding "utility" (bottom). The maximum indicates the next point where to evaluate the objective function

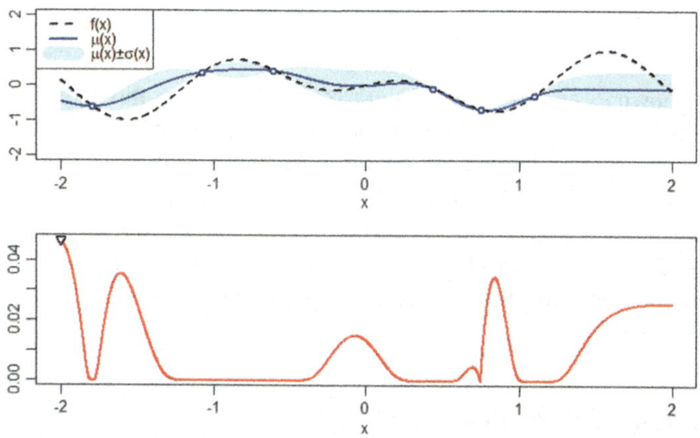

Fig. 1.11 How GP and acquisition function change after one more function evaluation, selected as in Fig. 1.10

One way to interpret a Gaussian process (GP) regression model is to think of it as defining a distribution over functions and with inference taking place directly in the space of functions. A Gaussian process is a collection of random variables, any finite number of which has a joint Gaussian distribution. A Gaussian process is completely specified by its mean function $\mu(x) = \mathbb{E}[f(x)]$ and covariance function $k(x, x') = \mathbb{E}[(f(x) - \mu(x))(f(x') - \mu(x'))]$. The covariance function $k(x, x')$ is also named *kernel* and amounts to impose a geometry to our decision space to Gaussian process depending on observations.

The Gaussian process is initially fitted on a set of available observations $D_{1:n}$ (STEP 1 of SMBO).

$$\mu_n(x) = \mathbb{E}[f(x)|D_{1:n}, x] = k(x, X_{1:n})\left[K(X_{1:n}, X_{1:n}) + \lambda^2 I\right]^{-1} y$$
$$\sigma_n^2(x) = k(x, x) - k(x, X_{1:n})\left[K(X_{1:n}, X_{1:n}) + \lambda^2 I\right]^{-1} k(X_{1:n}, x)$$

where $D_{1:n} = \{(x_i, y_i)\}_{i=1,\ldots,n}$ is the set of observation performed so far, the vector $\mathbf{y}$ gives the value of the function at the previous points, the covariance matrix K has entries $K_{ij} = k(x_i, x_j)$, with $i, j = 1, \ldots, n$, and $k(x, X_{1:n})$ is a n-dimensional vector with components $k_i = k(x, x_i)$.

GP also allows to build an "acquisition function" (such as the probability of improvement, PI, over the best value observed so far, introduced in the next subsection) which modulates the trade-off between exploration and exploitation and suggests the next promising candidate point x_{n+1}.

The GP is useful also because it provides an analytical expression for the update of mean and variance as a new function evaluation (x_{n+1}, y_{n+1}) arrives, by using the same formulas and simply replacing n with $n+1$.

In this chapter, the reference kernel for the GP is the squared exponential (SE):

$$k_{SE}(x, x') = e^{-\frac{\|x - x'^2\|}{2\ell^2}}$$

With ℓ known as *characteristic length-scale*. The role of this hyperparameter is to rescale any point x by $1/\ell$ before computing the kernel value. A large length scale implies long-range correlations, whereas a short length scale makes function values strongly correlated only if their respective inputs are very close to each other.

One should not see GP as one-size-fit-all recipe: many different kernels, as well as other models, have been suggested and are reviewed in Chap. 3. Also, many acquisition functions have been suggested and will be reviewed in Chap. 4.

1.2.3 Acquisition Function

Besides the surrogate model, the other main building block is the *utility function*, also named *acquisition function* or *infill criterion*.

Acquisition functions (STEP 2 of SMBO) are a critical part of the BO framework: many have been proposed, and new ones are being suggested to cope with new problems and take advantage of more computational power. Chapter 4 is devoted to this topic.

The first utility function was probability of improvement (PI) (Kushner 1964):

$$PI(x) = P\left(f(x) \le f(x^+)\right) = \phi\left(\frac{f(x^+) - \mu(x)}{\sigma(x)}\right)$$

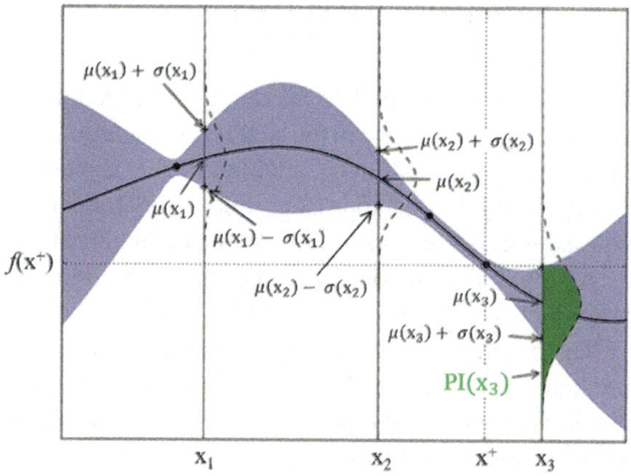

Fig. 1.12 A representation of probability of improvement

where $f(x^+)$ is the best value of the objective function observed so far, $\mu(x)$ and $\sigma(x)$ are mean and standard deviation of the probabilistic surrogate model, such as a GP, and $\phi(\cdot)$ is the probability density function.

This function will be analysed in detail in Chap. 4. This picture shows that the value of PI in x_3, given the best y value obtained in x^+, corresponds to the area depicted in green (Fig. 1.12).

GP, as surrogate model, has several advantages, which explain why it is widely used:

1. It provides a closed formula for mean and variance which can be analytically updated through new observations and takes care of measurement uncertainty and structural uncertainty (Figs. 1.4 and 1.5).
2. We can easily define an analytical acquisition function to deal with exploration—exploitation dilemma (this is a defining feature of BO, and its relationship with learning will be considered in Chap. 2).
3. The derivative of a GP is also a GP allowing to estimate in the training process not only the value of the function but also its gradients and hessians, given their, eventually noisy, observations: this enables to speed up the convergence reducing the variance in the GP regression (Chap. 3). This possibility is yet not really exploited in the available software being fraught with numerical instability, as we shall discuss in Chap. 3.

On the other hand, GP-based BO does not scale well with the number of decision variables (dimension of the design space). This is due, according to Kandasamy et al. (2015), to the statistical difficulty to scale up nonparametric regression in high dimensions and the computational challenge in maximizing the acquisition function. In Chap. 4, we shall see that many approaches have been proposed to mitigate these problems: random dropout and random embeddings of relevant search space dimensions (Wang et al. 2016; Chen et al. 2012) and structural assumptions such as the additive form of the objective function (Kandasamy et al. 2015).

1.3 Automated Machine Learning for Predictive Analytics

This section is devoted to an optimized machine learning pipeline that has been designed and developed in the context of research and customer projects and deployed in a number of situations (Candelieri 2017; Candelieri and Archetti 2018).

According to this approach, time-series forecasting is based on two consecutive phases: time-series clustering and artificial neural network (ANN) for regression.

As the loss function, we consider a measure of forecasting error, namely mean absolute percentage error (MAPE).

The problem consists of hyperparameter optimization where the number of ANN predictors to be learned in the second phase depends on the value of one of the hyperparameters of the first phase (i.e. number of clusters). In Fig. 1.13, we provide a representation of this problem:

Figure 1.14 summarizes the organization of the predictive pipeline. With respect to the previous picture, an emphasis on data used for training the overall system is also given.

Data was organized into a distinct time-series dataset $\mathcal{D} = \{x_1, \ldots, x_l\}$ consisting of l vectors, one for each day in the observation period, and where each vector x_i is a set of 24 ordered values, one for each hour of the day. The basic advantage of using time-series clustering as first stage is that it might allow for the identification of typical patterns within the time window of interest and, consequently, for splitting the dataset into subsets (i.e. clusters) which are then used in the second stage.

The clustering algorithm, kernel k-means, is a generalization of the standard k-means: it implicitly maps data from the input space, spanned by the original set of

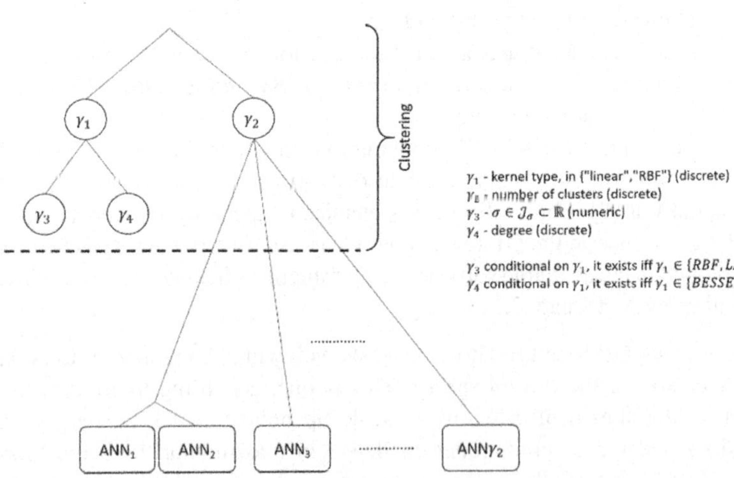

Fig. 1.13 Hyperparameter optimization problem for the predictive analytics pipeline considered. The number of ANN models to be trained in the second phase depends on the value of the hyperparameter defining the number of clusters in the first phase

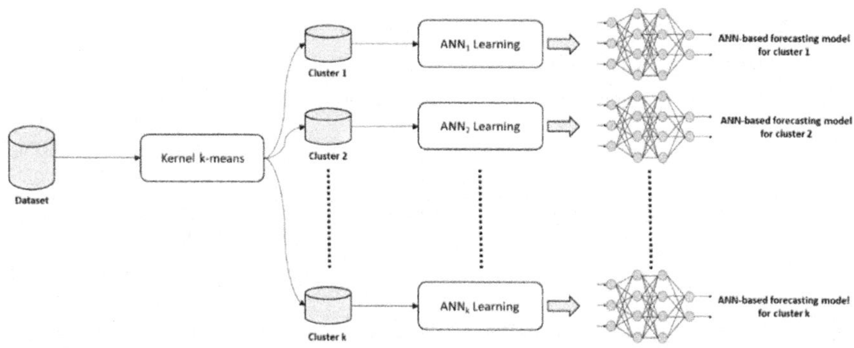

Fig. 1.14 A predictive analytics pipeline for time-series forecasting: first stage is clustering through kernel k-means; in the second stage, an ANN is learned on each cluster

data, to a higher-dimensional space, namely feature space. Therefore, kernel k-means can discover clusters that are nonlinearly separable in input space.

The second stage of the pipeline consists in training $\bar{k}$ different ANNs, one for each one of the $\bar{k}$ clusters resulting from the first stage. The first 6 values of the time series and the remaining $24 - 6 = 18$ are, respectively, the input and output neurons of every ANN. For each ANN, the hyperparameters reported in Table 1.2 are optimized via SMBO. The number of ANNs is not given a priori, and it depends on the number $\bar{k}$ of clusters identified from the first stage. The training process for the $\bar{k}$ ANNs is performed in parallel.

Tables 1.1 and 1.2 summarize, separately for the two stages, the hyperparameters

Table 1.1 Decision variables/hyperparameters for the clustering phase

Hyperparameter	Hyperparameter name	Type	Description
γ_1	$\bar{k}$	Integer	Number of clusters. Possible values are from 3 to 9
γ_2	Kernel type	Categorical	Type of kernel used in the kernel-based clustering. Possible kernels are: linear, spline, RBF, Laplace, Bessel, polynomial
γ_3	σ	Numeric, conditioned	Hyperparameter of the RBF, Laplace and Bessel kernels. Possible values are in $[10^{-5}, 10^{5}]$
γ_4	Degree	Integer, conditioned	Hyperparameter of the Bessel and polynomial kernels. Possible values are 2, 3 or 4

Table 1.2 Decision variables/hyperparameters for the ANN learning phase

Hyperparameter	Hyperparameter name	Type	Description
$\bar{\gamma}_1^k$	Hidden layers	Integer	Number of hidden layers in the artificial neural network. Possible values are 1, 2 or 3
$\bar{\gamma}_2^k$	Neurons in the hidden layer 1	Integer	Number of neurons in the hidden layer 1. Possible values are from 1 to 20
$\bar{\gamma}_3^k$	Neurons in the hidden layer 2	Integer, conditioned	Number of neurons in the hidden layer 2 (if $\bar{x}_1^k > 1$). Possible values are from 1 to 20
$\bar{\gamma}_4^k$	Neurons in the hidden layer 3	Integer, conditioned	Number of neurons in the hidden layer 3 (if $\bar{x}_1^k > 2$). Possible values are from 1 to 20
$\bar{\gamma}_5^k$	Algorithm	Categorical	Type of algorithm used to train the artificial neural network. Possible values are: backprop, rprop+, rprop-, sag, slr
$\bar{\gamma}_6^k$	Learning rate	Numeric, conditioned	Learning rate of the backprop algorithm. Possible values are in the range [0.1, 1.0]
$\bar{\gamma}_7^k$	Error function	Categorical	Function used to compute training error. Possible values are: sse and ce
$\bar{\gamma}_8^k$	Activation function	Categorical	Function used to compute the output of every neuron. Possible values are: logistic and tanh
$\bar{\gamma}_9^k$	Linear output	Logical	This hyperparameter defines whether to use a linear combination in the output layer of the artificial neural network or not. Possible values are TRUE or FALSE

of the predictive analytics pipelines (i.e. decision variables). According to the value of the hyperparameter γ_1 (i.e. the number of clusters), the number of hyperparameters in the second phase is consequently defined: it is γ_1 times 9. Since γ_1 is an integer ranging from 3 to 9, this means that the overall number of hyperparameters for the pipeline ranges from $4 + 3 \times 9 = 31$ to $4 + 9 \times 9 = 85$.

The performance measure is mean average percentage error (MAPE):

$$\text{MAPE} = \frac{1}{T}\sum_{t=1}^{T}\left|\frac{A_t - F_t}{A_t}\right|$$

where A_t and F_t are the actual and forecasted values, respectively, at time t, and T is the number of time steps predicted. Since we have a MAPE value for every time series and an ANN is trained for every cluster, MAPE is aggregated as:

$$\min_{x \in X}\left\{f(x) = \max_{k=1,\ldots,\bar{k}} \overline{\text{MAPE}_k}\right\}$$

This predictive pipeline was developed in R by using the following packages:

- "mlrMBO" for BO (described in Chap. 6), by setting a random forest as surrogate model and testing PI, EI and LCB as acquisition functions. The initial design has been sampled according to latin hypercube sampling (LHS) procedure
- "kernlab" for implementing the kernel k-means clustering
- "neuralnet" for implementing the artificial neural networks.

The approach has been validated on two different datasets, one public and related to energy consumption (https://archive.ics.uci.edu/ml/datasets/Appliances+energy+prediction), and one collected during a European project and related to urban water consumption. In both cases the first six actual values are sufficient to forecast, in one shot, the overall consumption pattern for the day. The resulting value of prediction error was very low (approximately 0.1%) for both the test cases considered.

This means that, when a new data comes (i.e. a vector of consumptions at the first six hours of the day) it is assigned to one of the clusters and, consequently, the corresponding predictor is selected and used to forecast. Thus, the overall pipeline is flexible to adapt the choice of predictor to the incoming data.

References

Adam, S.P., Alexandropoulos, S.A.N., Pardalos, P.M., Vrahatis, M.N.: No free lunch theorem: a review. In: Approximation and Optimization, pp. 57–82. Springer, Cham (2019)

Bengio, Y., Grandvalet, Y.: No unbiased estimator of the variance of k-fold cross-validation. J. Mach. Learning Res. **5**, 1089–1105 (2004, September)

Candelieri, A., Archetti, F.: Global optimization in machine learning: the design of a predictive analytics application. Soft Comput. 1–9 (2018). https://doi.org/10.1007/s00500-018-3597-8

Candelieri, A.: Clustering and support vector regression for water demand forecasting and anomaly detection. Water (Switzerland) **9** (2017). https://doi.org/10.3390/w9030224

Chen, B., Castro, R., Krause, A.: Joint optimization and variable selection of high-dimensional Gaussian processes (2012). arXiv preprint arXiv:1206.6396

Cozad, A., Sahinidis, N.V., Miller, D.C.: Learning surrogate models for simulation-based optimization. AIChE. **60**(6), 2211–2227 (2014)

Feurer, M., Springenberg, J.T., Klein, A., Blum, M., Eggensperger, K., Hutter, F.: Efficient and robust automated machine learning. In: Advances in Neural Information Processing Systems, pp. 2962–2970 (2015)

Florea, A.C., Andonie, R.: A dynamic early stopping criterion for random search in SVM hyperparameter optimization. In: IFIP International Conference on Artificial Intelligence Applications and Innovations, pp. 168–180. Springer, Cham (2018)

Frazier, P.I.: Bayesian optimization. In: Recent Advances in Optimization and Modeling of Contemporary Problems, pp. 255–278. INFORMS (2018)

Hutter, F., Kotthoff, L., Vanschoren, J.: Automatic machine learning: methods, systems, challenges. Challenges Mach. Learn. (2019)

Jiang, G., Wang, W.: Error estimation based on variance analysis of k-fold cross-validation. Pattern Recogn. **69**, 94–106 (2017)

Kandasamy, K., Schneider, J., Póczos, B.: High dimensional Bayesian optimisation and bandits via additive models. In: International Conference on Machine Learning, pp. 295–304 (2015, June)

Kotthoff, L., Thornton, C., Hoos, H.H., Hutter, F., Leyton-Brown, K.: Auto-WEKA 2.0: automatic model selection and hyperparameter optimization in WEKA. J. Mach. Learn. Res. **18**(1), 826–830 (2017)

Kushner, H.J.: A new method of locating the maximum point of an arbitrary multi-peak curve in the presence of noise. J. Basic Eng. **86**, 97–106 (1964)

Li, L., Jamieson, K., DeSalvo, G., Rostamizadeh, A., Talwalkar, A.: Hyperband: A novel bandit-based approach to hyperparameter optimization (2016). arXiv preprint arXiv:1603.06560

Nadeau, C., Bengio, Y.: Inference for the generalization error. In: Advances in neural information processing systems, pp. 307–313 (2000)

Shahriari, B., Swersky, K., Wang, Z., Adams, R.P., De Freitas, N.: Taking the human out of the loop: A review of Bayesian optimization. In: Proceedings of the IEEE, pp. 148–175 (2016)

Strubell, E., Ganesh, A., McCallum, A.: Energy and policy considerations for deep learning in NLP (2019). arXiv preprint arXiv:1906.02243

Thornton, C., Hutter, F., Hoos, H. H., Leyton-Brown, K.: Auto-WEKA: combined selection and hyperparameter optimization of classification algorithms. In: Proceedings of the 19th ACM SIGKDD international conference on Knowledge discovery and data mining, pp. 847–855. ACM (2013, August)

Wang, Z., Hutter, F., Zoghi, M., Matheson, D., de Freitas, N.: Bayesian optimization in a billion dimensions via random embeddings. J. Artif. Intell. Res. **55**, 361–387 (2016)

Wolpert, D.H., William, G.M.: No Free Lunch Theorems for Search, Vol. 10. Technical Report SFI-TR-95-02-010, Santa Fe Institute (1995)

Wolpert, D.H.: The supervised learning no free-lunch theorems. In: Soft Computing and Industry, pp. 25–42. Springer, London (2002)

Zhu, D., Linke, N.M., Benedetti, M., Landsman, K.A., Nguyen, N.H., Alderete, C.H., Perdomo-Ortiz, A., Korda, A., Garfoot, A., Brecque, C., Egan, L., Perdomo, O., Monroe, C.: Training of quantum circuits on a hybrid quantum computer (2018). arXiv preprint arXiv:1812.08862

Chapter 2
From Global Optimization to Optimal Learning

What is the relation between finding the global minimum of the function below and the learning paradigm (Fig. 2.1)? What learning models have in common with global optimization methods? Outlining possible answers and linking them to other parts of the book are the objective of this chapter.

Some references are in order to offer a framework for the development of global optimization. The two volumes (Dixon and Szegö 1975, 1978), albeit predated by a number of pioneering papers, might be regarded as the first concerted effort to raise the issue of global optimization as a new topic in the optimization community and have been highly influential for the development of the subject. Other volumes, of general interest, which drove the development of the field of global optimization are

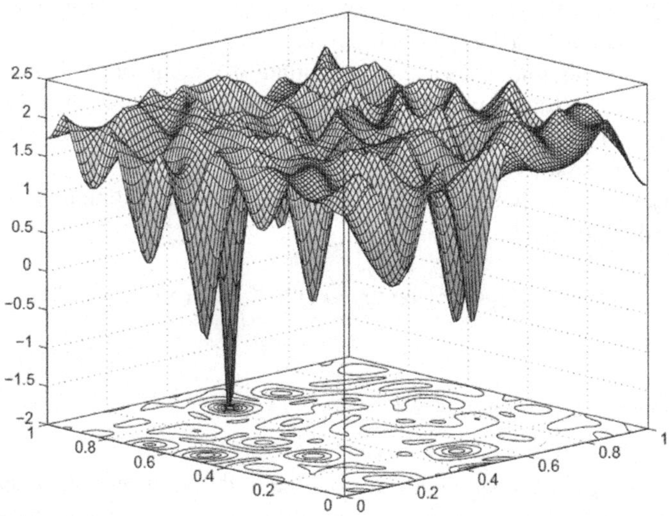

Fig. 2.1 A multi-extremal two-dimensional objective function. (*Source* Sergeyev and Kvasov 2017)

F. Archetti and A. Candelieri, *Bayesian Optimization and Data Science*,
SpringerBriefs in Optimization, https://doi.org/10.1007/978-3-030-24494-1_2

Pardalos and Romeijn (2013), Pintér (2006), Powell and Ryzhov (2012), Locatelli and Schoen (2013).

2.1 A Priori Analysis of Global Optimization Strategies

The aim of this section is not to deal with specific algorithms, but to highlight the theoretical framework underlying the design of deterministic algorithms and to assess the computational complexity of global optimization problems using "space-covering" techniques.

Let X be a compact set in the N-dimensional Euclidean space R^N. We define the global optimization problem (GOP) as finding the couple x^*, f^* such that:

$$f^* = f^*(x^*) \leq f(x) \forall x \in X$$

where $f : X \rightarrow R$, $f \in C(X)$.

The very definition of the problem is not well-posed in the mathematical sense. Indeed, it is possible to choose multi-extremal functions, which are close to one another as we want while their global extrema are far apart. In other words, the global minimizer x^* does not depend continuously on the data of the problem. This problem, which generates numerical instability, haunts also very recent approaches which propose information theoretic-based acquisition functions, which will be considered in Chap. 4.

On the other hand, the value f^* depends continuously on the data of the problem, so we shall generally restrict the search to f^* for which we can show that, for continuous functions, the GOP is well-posed.

Still it is not possible under the only continuity assumption, to bound the error in the approximate solution: given the dataset $D_{1:n}$, it is possible to build a smooth function g, as shown in Fig. 1.4, interpolating the point in the dataset whose global minimum g^* is as far as we wish from f^*.

The easiest way to bound the error in the approximate solution is to impose a Lipschitz condition on f:

$$f \in L_{\ell,\rho}(X) = \{g : |g(x) - g(y)| \leq \ell\rho(x, y) \forall x, y \in X\}$$

where

ℓ is a known positive constant and
ρ is a continuous distance on X.

Under the Lipschitz condition, it is possible to design deterministic algorithms (space-covering techniques) which provide an estimate of f^* whose distance from the optimum can be deterministically controlled.

The design of these algorithms is based on three concepts (Archetti and Betrò 1978):

- *Strategy*: for any n, a strategy is a vector-valued function mapping $L_{\ell,\rho}(X)$ into $X^n = X \times \cdots \times X$, ($n$-points strategies will be, in the following of this Section, be indicated by S_n and the n-tuple they associate to f by $S_n(f) = (x_1(f), \ldots, x_n(f))$. If the strategy is constant in $L_{\ell,\rho}(X)$, i.e., it maps $L_{\ell,\rho}(X)$, into a point of X^n, then it is termed passive. If the strategy depends on f so that $x_j(f), j = 1,\ldots,$n depends on x_l and $f(x_l)$, $l = 1, \ldots, j-1$, then it is termed sequential.
- *Accuracy* of S_n, $S_n(f) = (x_1(f), \ldots, x_n(f))$, for $f \in L_{\ell,\rho}(X)$, is given by $A(S_n, f) = \min_{i=1,\ldots,n} f(x_i) - f^*$. The guaranteed accuracy of S_n in $L_{\ell,\rho}(X)$ is A $(S_n) = \text{Sup}_{f \in L_{\ell,\rho}(K)} A(S_n, f)$
- *Optimality*: S_n^* is A-optimal if $A\left(S_n^*\right) = Inf_{S_n} A(S_n)$. Given $\delta > 0$ if there exists n^* and S_{n^*} such that $A(S_{n^*}) \leq \delta$, while no strategy S_n exists such that $A(S_n) \leq \delta$, $n < n^*$, then S_{n^*} is n-optimal.

In general, the problem of finding n-optimal strategies can be reduced to a space-covering problem. Given a uniform grid in X with mesh size h, it can be shown that a value $\bar{h}$ exists such that, for $h < \bar{h}$; $n^*(h)$, the cardinality of the discrete optimal covering, is equal to n^*. This result implies that A- and n- optimal strategies require, for a given guaranteed accuracy δ in the approximation to f^*, a number of function evaluations which increases exponentially in the number of independent variables of the problem.

Indeed, the space-covering problem is NP-complete and its optimization version as integer linear program is NP-hardness. We could sum up this analysis of the deterministic approach by the term "computational intractability" of GOP.

The equivalence with an integer programming problem spawned a number of branch and bound-based approaches to the GOP, in particular probabilistic branch and bound (Zhigljavsky 1990; Norkin et al. 1998; Zabinsky et al. 2011).

Still one could think that sequential strategies can perform better than passive ones: this is not the case, at least in terms of guaranteed accuracy.

We remark that the above results are "worst case" concerned with the guaranteed accuracy in $L_{\ell,\rho}(X)$: for most functions in $L_{\ell,\rho}(X)$, a sequential strategy performs better than passive ones, and for this reason, an important body of research has been devoted to develop a sequential algorithm with early results due to (Evtushenko 1971; Shubert 1972; Strongin 1978). Recent results brought to the forefront the design of new computational frameworks enabling algorithms with vastly improved numerical performances (Sergeyev and Kvasov 2017) which shall be considered in the next section.

2.2 Lipschitz Global Optimization (LGO)

As we have seen to obtain estimates of the global solution, some quantitative assumptions on the objective functions and constraints are required. A powerful as well as natural assumption is that they have bounded rate of variation. This assumption is justified because, at least in technical systems, the rate of change is typically limited

(Strongin 1978, 1992). In many cases, the value of the Lipschitz constant L or an upper bound can be dictated by the physical laws of the underlying process. This technical condition is mathematically translated into the Lipschitz condition, which allows, at least in one dimension, a simple geometric interpretation establishing a lower bound as shown in Fig. 2.2.

$$\varphi_i(x) = \max\{z_{i-1} - L(x - x_{i-1}), z_i + L(x - x_i)\}$$

The function $\varphi_i(x)$ over $\left[x_{i-1}, x_i\right]$ is therefore a lower bound of $f(x)$ over this interval. The value where the minimum is attained is a reasonable suggestion for the next evaluation point. In a sense, $\varphi_i(x)$ gives a worst case estimate of the uncertainty of the objective function.

This is the basic idea upon which various modifications have been proposed, collectively labelled "geometric algorithms".

There is a strong relation between these geometric approaches and the informational statistical approaches (Strongin 1978).

These approaches can be extended to the multidimensional case, bearing in mind that finding the minimum of $\varphi_i(x)$ is a multi-extremal problem, and in the worst case, the number of evaluations is anyway exponential.

The restriction of the class of the objective function to the differentiable functions with the first derivative satisfying a Lipschitz condition allows to develop efficient geometric LGO methods: $\theta_i(x)$ is the new lower bound, built on smooth piecewise quadratic auxiliary functions $\psi_i(x)$, which speeds the searching process up (Fig. 2.3).

One of the first methods, for solving the one-dimensional LGO is due to Shubert (1972). A significant advancement in LGO was proposed in Jones et al. 1993 under the name of Dividing RECTangles (DIRECT) in which several estimates of the Lipschitz constant are used at each iteration. Also, Strongin's algorithm provides an adaptive estimation of the Lipschitz constant during the global search.

DIRECT tries to balance global and local information, exploration–exploitation in BO jargon, but it is basically driven by local information and therefore towards exploitation. A new geometrical algorithm MultL, inspired by DIRECT, has been

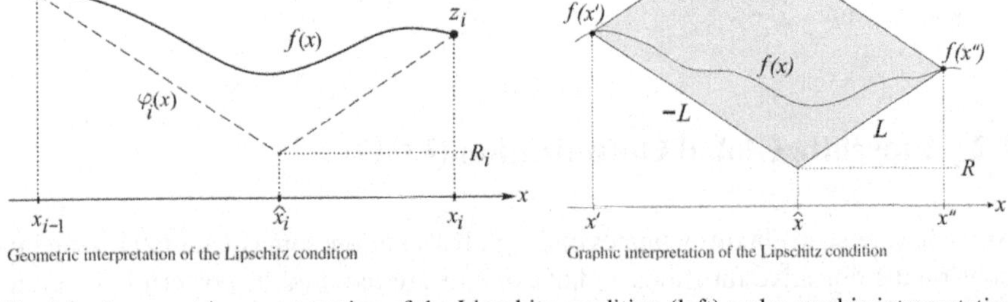

Fig. 2.2 A geometric representation of the Lipschitz condition (left) and a graphic interpretation of the Lipschitz condition (right). (*Source* Sergeyev and Kvasov 2017)

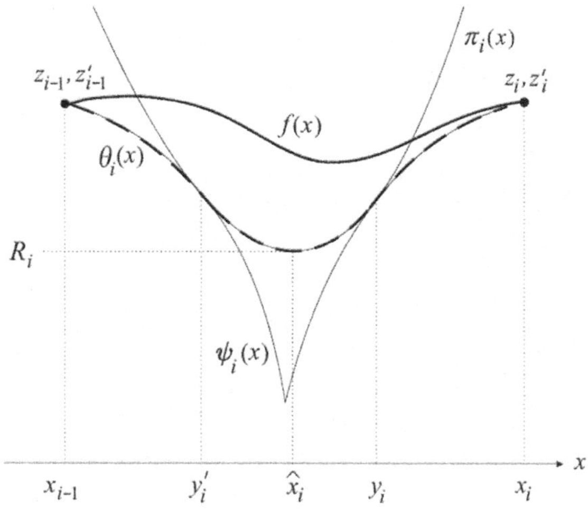

Fig. 2.3 Smooth piecewise quadratic auxiliary function $\theta_i(x)$ (dashed line) for the objective function $f(x)$ (thick line) with the Lipschitz first derivative over $[x_{i-1}, x_i]$. (*Source* Sergeyev and Kvasov 2017)

proposed in Sergeyev and Kvasov (2006), which can be integrated with the efficient diagonal partition strategy, discussed in Sergeyev and Kvasov (2017) to yield what is widely regarded as the best LGO algorithm to solve hard multidimensional black-box optimization problems.

The main question is how to obtain the value of L. Many approaches have been proposed for which the reader can address (Sergeyev and Kvasov 2017): the most interesting one is an adaptive estimation of L which can yield either global estimate of the Lipschitz constant or local estimate valid for some region.

Similar arguments have been used in Malherbe and Vayatis (2017), which propose a method to design sequential strategies for efficient optimization of a black-box function under the only assumption that it has a finite Lipschitz constant.

Another critical issue is the "curse" of dimensionality: to mitigate it (Strongin 1992) uses Peano's space-filling curves and solves the problem in one dimension, albeit with an increase in the Lipschitz constant. Besides the theoretical and computational relevance of Lipschitz optimization by itself, Lipschitz and Bayesian optimization can be hybridized quite naturally incorporating the Lipschitz inequality bound in the acquisition function in order to prevent BO from considering x values which cannot be globally optimal and prevent exploration on unnecessary regions of the search space. The use of the Lipschitz condition has been first proposed in Jalali et al. (2013) and more recently in Ahmed et al. (2018). The following figures show, respectively, (top-left) the GP model after 5 iterations, with its mean and variance; (top-right) the excluded regions according to the (known) L and (bottom) the combination of the two regions, showing how LBO helps cut off regions where the posterior variance is high but can be excluded according to Lipschitz constant (Fig. 2.4).

The Lipschitz condition has been also used with BO in the context of "safe optimization"; when the objective function has a safety threshold which cannot be

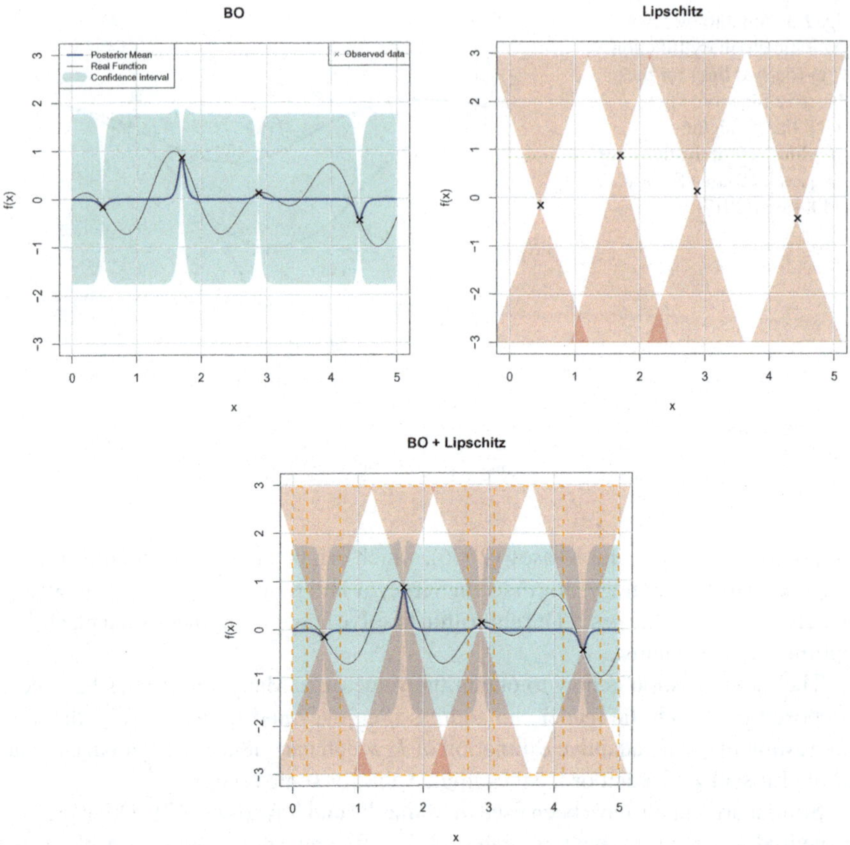

Fig. 2.4 A GP conditioned on four observations, with mean and variance (top-left), excluded search areas due to the Lipschitz condition (top-right) and intersection between GP variances and excluded zone (bottom)

crossed, and the acquisition function must suggest points in the safe region (Sergeyev et al. 2019). This subject will be taken over in Chap. 5.

2.3 Random Search

Random Search (RS) methods have been first considered in Brooks (Brooks 1958) and developed into a coherent general framework by (Rastrigin 1963). A masterful analysis of the statistical issues associated with RS is contained in Zhigljavsky (1985). Its use has been recently advocated for hyperparameter optimization (Bergstra and Bengio 2012) and reinforcement learning (Mania et al. 2018). Some topics are considered also in more recent publications (Zhigljavsky and Zilinskas 2007; Zabinsky 2011; Zilinskas and Zhigljavsky 2016; Dodge et al. 2017).

According to the general RS framework, a sequence of random points $x_1, x_2, \ldots$ is generated according to a *pdf* which might in general depend on previous points x_j and the values $f(x_j)$.

Within this generic framework, different computational strategies can be embedded. In the basic Random Search points, $x_1, x_2, \ldots$ are sampled from a uniform distribution in the search space X and the sequence is stopped after a prefixed number of function evaluations or some statistical test is satisfied. (Archetti 1975.

Another specialization is when P_j is uniform distribution in the level set $A_j = \left(x s.t. f(x) \leq f\left(x_j^+\right)\right)$, where $f\left(x_j^+\right)$ the "best seen" among the sample $x_1, x_2, \ldots, x_j$.

The resulting algorithm, called pure adaptive search (Zhiglavsky 1985; Zabinsky and Smith 1992), has theoretically very good convergence, but its actual computational efficiency is very poor because generating random points uniformly in A_j is itself computationally very expensive. Other contributions to RS for global optimization are Zhigljavsky and Chekmasov (1996), Wang (2004), Missov and Ermakov (2009) and Zabinsky (2015).

Methods based on random sampling can be coupled effectively with local searches according to the following pattern:

(i) Draw q points $\{x_j \in X\}$, $j = 1, \ldots, q$ from a uniform distribution in X and compute $f(x_j)$.
(ii) Select the "most promising points" and start from them a local optimization routine obtaining a least value f_c.
(iii) Test whether f_c is the global minimum of f in X.

The above cycle is repeated until the test is satisfied. In the simplest such algorithm, the point yielding the least sampled value is used in (ii) as the starting point, and the test in phase (iii) is satisfied if no improvement over f_c observed in one or more further executions of phase (i).

This simple test relies upon the fact that, as the sample size increases, the probability of not improving over f_c decreases unless f_c is the global minimum.

In this method, no more than one local optimization is performed in the region of attraction of a minimum: this good feature, unfortunately, is obtained at the cost of wasting most of the information contained in the sample.

Effective algorithms require a more balanced compromise between the conflicting goals of making good use of the available information and of reducing the risk of converging to a local minimum already found (exploration vs exploitation): an effective way to accomplish this is through cluster analysis.

The third step is clearly the critical part of these algorithms: it is very difficult to evaluate the probability of the result being exact or, more generally, the probability that some meaningful index of the error does not exceed a prefixed level. Some proposals have been suggested to frame the third step into correct statistical terms and will be discussed in detail later in Sect. 2.3.3.

2.3.1 General Properties of Uniform Sampling

The uniform sampling of f in X is meant to provide a sample, an "exploration basis" for the inferential processes to be carried out on in the following stages of the algorithm, rather than to provide directly an approximation to f^*. Still, a "per se" analysis of uniform sampling can be performed, albeit only to some extent and in mainly negative terms. Let $X_0 \subseteq X$ be such that $m(X_0)/m(X) = \alpha$, where $m(.)$ denotes the volume: the probability $P(X_0; q)$ that at least one point in the sample $\{x_j\}, j = 1 \ldots .q$ will belong to X_0 is given by

$$P(X_0; q) = 1 - (1 - \alpha)^q$$

Now, let X_0 be a neighbourhood of x^*: one could derive, given a value α) and a probability level $\bar{P}$, a sample size $\bar{q} = \log(1 - \bar{P})/\log(1 - \alpha)$ and assume $f^*_{\bar{q}} = \min_{j=1,\bar{q}} f(x_j)$ as an approximation to f^*.

Nothing can be said, for finite values of q and without specific assumptions about f, about the probability that the error $f^*_{\bar{q}} - f^*$ exceeds a prefixed value, nor can we assure that the value $f^*_{\bar{q}}$ has been achieved in X_0.

2.3.2 Cluster Analysis

Multi-start methods, based on a clever combination of RS and cluster analysis, can be regarded as effective tools for the numerical solution of GOP when derivatives are available and function evaluations are not too expensive. It is important to remark that the derivative requirement reduces the usefulness of random multi-start in hyper-parameter optimization and in general black-box or simulation-based optimization.

The relevance of cluster analysis to the numerical solution of global optimization problems, first stressed in Torn (1978), has been subsequently substantiated by a number of implementations. A successful exploitation of clustering methods is given in Boender et al. (1982), Locatelli and Schoen (1999), Schoen (1998), where random and quasi-random linkage methods are considered. Very recent contributions to this topic are Zilinskas et al. (2019) and Bagattini et al. (2019).

2.3.3 Stopping Rules

A basic fact about global optimization algorithms is that the global part of a successful algorithm, actually the probabilistic part of it, is connected not with the numerical approximation of the global optimum, which is effectively performed by local searches, but rather with the control of these local searches and the decision whether a local minimum can be accepted as the global one.

A theoretical analysis of random multi-start has been initiated by Zielinski (1981), who constructed a multi-nomial distribution such that a set of local extrema obtained by performing a number of local searches from uniformly distributed starting points can be interpreted as a sample drawn from this multi-nomial distribution. Therefore, a posteriori probability can be computed that another local search will lead to the identification of a new local minimum and that can be used to determine optimal Bayesian stopping rules. Its results were carried over by Betrò (1984), Betrò and Rotondi (1984), Schoen (1998). A dynamic early stopping rule for Random Search has been recently proposed in Florea and Andoine (2018).

Another approach employs tools from order statistics to obtain the distribution of $f_c = \min_{j=1,q} f_j$, where $\{x_j\} = 1 \ldots q$ be uniformly distributed points in X (Clough 1969; Zhigljavsky and Zilinskas 2007). Unfortunately, an exceedingly large sample is required for this approach to be of significant practical value.

An interesting hybridization of Random Search and BO has been proposed recently in hyperband (Li et al. 2016) for the problem of hyperparameter optimization. The approach focuses on speeding up Random Search through adaptive resource allocation and early stopping. Hyperparameter optimization is defined as a pure-exploration infinite-armed bandit problem where a resource, like data samples, is allocated to randomly sampled configurations.

The basic idea is very simple. Given N hyperparameter configurations to evaluate, we devote initially 1 unit of budget to each configuration, then 2 units to the best half and continue halving configurations and doubling individual allocations until we have budget left.

Reportedly (Li et al. 2016), Hyperband outperforms BO: still it shares with RS the problem of inefficient use of previous information. To overcome this drawback, Falkner et al. (2017) propose the idea to integrate hyperband with a Tree Parzen Estimator (TPE) model, described in Chap. 3.

The random sampling of configurations is replaced by a model-based search: when the number of iterations for each configuration is reached, the halving procedure is carried out.

In Wang et al.(2018), the idea of combining hyperband and BO has been further carried out for hyperparameter optimization in deep learning networks.

Another important connection is between Random Search and heuristics like simulated annealing (SA) and genetic algorithm (GA): it is shown, in Zhigljavsky and Zilinskas (2007), that any global RS can be formulated as a Markovian algorithm based on sampling from a Markov chain. One such algorithm is the SA which uses a previously sampled point. One would expect that using even more information might bring to the search more efficient. This can indeed be accomplished by developing population-based Markovian algorithms, including popular GAs, that transition from one group of points (current generation) to another group (next generation) by some probability rules (Zhigljavsky and Zilinskas 2007, Sect. 2.3.4 e 3.5).

If generations have the same size N, the search algorithms are still Markovian in the product space X^N, i.e. the underlying Markov chain is of the Nth order.

Evolutionary algorithms are increasingly used in stochastic and global optimization. We are not getting in a specific analysis which would be outside the scope of this book; still few things have to be remarked. Evolutionary or genetic algorithms do not handle explicitly the exploration–exploitation dilemma. Given the population nature, they are biased towards exploration which contributes to the global property but also to a slow convergence (Sergeyev et al. 2018).

A strain of GAs has incorporated the aspects of Bayesian methods under the name of Bayesian Optimization Algorithm (BOA) (Pelikan et al. 1999) and estimation distribution algorithm (EDA) (Hauschild and Pelikan 2011), which uses Bayesian networks to estimate the joint probability of promising solutions (strings), and covariance weighted adaptation (CWA) which is a kind of "kernelization" of a genetic algorithm (Hansen 2016). Another meta-heuristic, particle swarm optimization (PSO), has been largely investigated for solving GOP (Parsopoulos and Vrahatis 2002). Close to the spirit of BO, the use of GP to assist an evolutionary method was proposed in Ulmer et al. (2003).

2.4 Bandits, Active Learning and Bayesian Optimization

A very recent and stimulating contribution to the general issue of learning sequential strategies is given in a DeepMind's technical report (Ortega et al. 2019). In this section, we focus on the more specific issue of Active Learning (AL), a strategy to select the most informative query point for predicting a varying environment, as, for instance, the objective function $f(x)$ modelled by a GP. GPs are the setting where most of the AL researches have been focused. A relevant presentation is provided in "*Bandits, Global Optimization, Active Learning, and Bayesian Reinforcement Learning—understanding the common ground*", *Autonomous Learning Summer School, Leipzig, Sep 2014, by Marc Toussaint.* More recently, Ling et al. (2016) move one step further towards unifying BO, AL and multi-armed bandits (MAB). MAB is the archetypal problem in online optimal learning. One-arm bandit refers to a slot machine operated by a lever (arm); multi armed refers to a collection of slot machines M_i, $i = 1,\ldots, n$ each with different random reward $r \sim P(r_i, \theta_i)$ for each machine.

We have money to play N times, and we should like to know the expected value for each machine $\bar{r}_i$; but the only way to get the actual reward is to put money in M_i, pull the lever and observe the outcome (i.e. the realization of r_i).

MAB is often used to model the *A*/*B* testing problem and recommender systems, in which we are running an online advertisement system and we can choose, dynamically, among different advertisements (e.g. banners) on the basis of their profitability. The critical point in MAB problem is, as in BO, balancing exploration (pulling an arm that might turn out to be a big winner) and exploitation (putting money on that arm which resulted in the best winning obtained so far). Indeed, the choice of a machine has two effects:

- increase one's knowledge
- increase one's reward.

Learning policies in MAB face the exploration versus exploitation dilemma, i.e. the search for a balance between exploring the environment to find profitable actions while taking the empirically best action as often as possible (Auer et al. 2002).

The problem is to find a policy that maximizes the expected long-term return: $\sum_t \gamma_t r_t$ where γ_t is the discount factor, defining the relative weights between immediate and long-term rewards.

The player has a number of actions $a_t \in (1, \ldots, n)$ representing the choice of a machine to play at time *t;* let $r_t \in \mathbb{R}$ be the outcome with mean $\bar{y}$

A policy or strategy maps all the history to the next machine to play:

$$\pi : \left[(a_1, r_1), \ldots, (a_{t-1}, r_{t-1})\right] \to a_t$$

This can be interpreted as our usual dataset $D_{1:n} = \{(x_i, y_i)\}_{i=1,\ldots,n}$ where $a_t = x_i$ and $r_t = y_i$.

Continuous global optimization can be seen as an infinite-arm bandit, in which the actions are the variable *x*, the expected reward is modelled by the probabilistic surrogate model, and the observed reward is the value of the objective function *f(x)*.

This way BO can be seen naturally also as an instance of active learning: in standard ML, the dataset is given; in AL, the sequential model-based optimizing agent sequentially decides on each *x*, i.e. where to collect data.

We are concerned with GP-based optimization in the MAB setting, where the reward is sampled from a GP distribution. To demonstrate convergence properties, it is more convenient to work with the concept of "regret" instead of "reward".

For the candidate action x_t at round t, we incur instantaneous regret $r_t = f(x_t) - f(x^*)$. The cumulative regret R_T after T rounds is the sum of instantaneous regrets: $R_T = \sum_{t=1}^{T} r_t$. A desirable asymptotic property of an algorithm is to be no-regret: $\lim_{T\to\infty} \frac{R_T}{T} = 0$. Note that neither r_t nor R_T are ever revealed to the algorithm. Bounds on the average regret $\frac{R_T}{T}$ translate to convergence rates for GP-based BO: the value $\max_{x_t \le T} f(x_t)$ in the first T rounds is no further from $f(x^*)$ than the average of the $f(x_t)$ samples.

An optimal policy π is the no-regret policy for the MAB and is offered by GP-upper confidence bounding (GP-UCB), which computed by solving the auxiliary problem:

$$x_{t+1} = \underset{x \in X}{\operatorname{argmax}}\, \mu_t(x) + \xi_t \sigma_t(x)$$

where $\mu_t(x)$ and $\sigma_t(x)$ are the mean and standard deviation of the GP at step t.

A proof of convergence of UCB, along with condition on ξ_t, is given in Srinivas et al. (2009). It is easy to implement and became, along with EI, the most used acquisition function in BO and will be discussed in Chap. 4.

The bandit problem is an archetype for

- sequential decision-making
- decisions that influence knowledge as well as rewards/states
- exploration–exploitation
- online learning.

which are the same features characterizing global optimization, active learning and reinforcement learning.

Another unifying perspective is Reinforcement Learning (RL). RL needs a different set-up, namely a Markov Decision Process (MDP), specified by *states*, *actions*, *transitions* and *rewards*. The definition of a *value function* is also needed, providing a value of a state $s_t = s$ at time step t, given the policy π. This value is denoted by $V^{\pi}(s)$ and is computed as the expected return of costs (or rewards) when starting in state s and by following, sequentially, the actions suggested by π:

$$V^{\pi}(s) = \mathbb{E}_{\pi}\left[\sum_{k=0}^{\infty} \gamma^k \cdot r_{t+k} | s_t = s\right]$$

where $\mathbb{E}_{\pi}$ is the expected value given by following policy π, and $\gamma \in [0, 1]$ is a *discount factor* that is used to balance current and future rewards. When γ is small the approach is said to be "myopic", which means that it is only concerned about short-term rewards, while when γ is large long-term rewards become also important. Optimizing $V^{\pi}(s)$ is equivalent to solving the dynamic programming equation and runs into the "curse of dimensionality". Moreover, it requires the knowledge of the complete structure of the MDP.

Most of the relevant real-life problems are to be solved in conditions of partial knowledge, and agents interacting with the environment often need to compute policies with few or no data, where performing an action and observing its outcome is the only way to obtain a better estimate of its value and to increase the knowledge about the system. Approximate dynamic programming (ADP) is largely overlapping with RL. Both are learning enabled policy optimization methods and can use Bayesian learning.

BO and RL can be knitted together in different ways. One possibility is related to how RL can improve BO: in Chap. 4 will be presented a mechanism to dynamically choose, from a portfolio of alternatives, the most promising acquisition function. Other possibilities are related to how BO can improve RL:

- BO for hyperparameter optimization of RL algorithm (Barsce et al. 2017), such as the number of epochs, the number of episodes, the value of ε in the case of ε-greedy policy, the learning rate and the discount factor.
- A GP approximates the reward function and a mechanism based on it, analogously to the non-myopic acquisition function in BO, is used to select the next action in a RL algorithm (Engel et al. 2003).
- BO for optimizing a "parametrized" policy: more precisely, the optimal value of the policy's parameters is identified through BO considering the reward cumulated

along the "trajectories" along with episodes as a black-box function (Wilson et al. 2014).

In this paper, the probability of observing a trajectory $\tau = (s_0, a_0, \ldots, a_{T-1}, s_T)$, i.e. a sequence of states and actions, given the agent follows a policy with parameters θ is:

$$P(\tau|\theta) = P_0(s_0)\prod_{t=1}^{T} P(s_t|s_{t-1}, a_{t-1})P_\pi(a_{t-1}|s_{t-1}, \theta)$$

where every trajectory τ begins in an initial state s_0 and terminates after at most T steps. A widely used policy is soft-max that in this case is parametrized as follows:

$$P(a|s) = \frac{e^{(\theta_a \cdot \phi(s))}}{\sum_{y \in A} e^{(\theta_y \cdot \phi(s))}}$$

with $\phi(s)$ a feature function mapping a state s into a vector.

The value of a trajectory is computed as the sum of the reward received by interacting with the system:

$$\bar{R}(\tau) = \sum_{t=0}^{T} R(s_t, a_t, s_{t+1})$$

The objective function is the expected return and computed as:

$$\eta(\theta) = \int \bar{R}(\tau)P(\tau|\theta)d\tau$$

The goal of "policy search" is to identify the policy's parameters θ that maximizes expectation:

$$\theta^* = \arg\max_{\theta} \eta(\theta)$$

The values of $\eta(\theta)$ are computed through Monte Carlo simulation.

The main features of this approach are a new kernel, which measures similarity between trajectories and a new GP mean function which uses rewards and learned transitions.

The exploration–exploitation dilemma, and the temporal trade-off over uncertain rewards, is also a central topic in cognitive science. We are not getting in a specific analysis, which would be outside the scope of this book, but we can quote at least some recent results: Schulz (2012) which proposes a Bayesian mechanism that can generalize beyond observed outcomes driving towards the formation of inductive beliefs about novel options (Gershman 2018a), where GP-based BO with UCB is suggested as a combined model of generalization and exploitation and (Gershman

2018b) which considers random exploration, by inserting randomness in the choice behaviour versus directed exploration driving choices towards uncertain options. This behavioural problem can also be modelled as a spatially correlated multi-armed bandit task in which rewards are distributed on a grid characterized by spatial correlation (high rewards tend to cluster together) (Gopnik et al. 2017, Schulz et al. 2018a, b), Wu et al. (2018) analyse how humans search for rewards under limited search horizons, where the spatial correlation of rewards provides traction for generalization.

The issue of GP and cognition is also studied in a recent survey by Schulz et al. (2018a).

The problem usually considered is function learning: Schulz et al. (2015) and Schulz et al. (2016) use different kernels for different degrees of smoothness. Wilson et al. (2015) propose a kernel learning framework "the human kernel" to account for the inductive biases of human learners. Another GP-based approach is proposed in Griffiths et al. (2009), where the equivalence between Bayesian linear regression and GP provides a unifying framework for function learning.

A more direct approach has been taken in Borji and Itti (2013) who analyse how the "problem-solving" behaviour of humans in the optimization of a black-box function compares to algorithms. The experimental results show that humans outperform most optimization algorithms and that GP-based BO is the best predictor of human behaviour. A recent paper addresses the study of cognitive model priors for predicting human decisions (Peterson et al. 2019). This paper moves from the framework of prospect theory and a remark that noisy human behaviour requires far larger datasets than in the usual studies; the proposed approach is then carried out testing the prediction accuracy of several machine learning algorithms for different cognitive models and a very large dataset containing more than 200 k human judgements over 13 k decision problems.

References

Ahmed, M.O., Vaswani, S., Schmidt, M.: Combining Bayesian Optimization and Lipschitz Optimization (2018). arXiv preprint arXiv:1810.04336

Archetti, F., Betrò, B.: A priori analysis of deterministic strategies. In: Dixon L., Szego G.P. (eds.) Towards Global Optimisation, vol. 2. North Holland (1978)

Archetti, F.: A sampling technique for global optimisation. In: Dixon L., Szego G.P. (eds.) Towards Global Optimisation, vol. 1. North Holland (1975)

Archetti, F.: A stopping criterion for global optimization algorithms. Quaderni del Dipartimento di Ricerca Operativa e Scienze Statistiche A-61 (1979)

Auer, P., Cesa-Bianchi, N., Fischer, P.: Finite-time analysis of the multiarmed bandit problem. Mach. Learn. **47**(2–3), 235–256 (2002)

Bagattini, F., Schoen, F., Tigli, L.: Clustering methods for large scale geometrical global optimization. Optim. Methods Softw. 1–24 (2019)

Barsce, J.C., Palombarini, J.A., & Martínez, E.C.: Towards autonomous reinforcement learning: automatic setting of hyper-parameters using Bayesian optimization. In: 2017 XLIII Latin American Computer Conference (CLEI), pp. 1–9. IEEE (2017, September)

Bergstra, J., Bengio, Y.: Random search for hyper-parameter optimization. J. Mach. Learn. Res. **13**, 281–305 (2012)

Betro, B.: Bayesian testing of nonparametric hypotheses and its application to global optimization. J. Optim. Theory Appl. **42**(1), 31–50 (1984)

Betrò, B., Rotondi, R.: A Bayesian algorithm for global optimization. Ann. Oper. Res. **1**(2), 111–128 (1984)

Boender, C.G.E., Kan, A.H.G.R., Timmer, G.T.: A stochastic method for global optimization. Math. Program. **22**, 125–140 (1982)

Borji, A., Itti, L.: Bayesian optimization explains human active search. Adv. Neural Inf. Process. Syst. **26**(NIPS2013), 55–63 (2013)

Brooks, S.H.: A discussion of random methods for seeking maxima. Oper. Res. **6**, 244–251 (1958). https://doi.org/10.1287/opre.6.2.244

Clough, D.J.: An asymptotic extreme value sampling theory for estimation of global maximum. CORS J. 102–115 (1969)

Dixon, L.C.W., Szegö, G.P.: Towards Global Optimization 1. Dixon, L.C.W., Szegö, G.P. eds. North Holland (1975)

Dixon, L.C.W., Szegö, G.P.: Towards Global Optimization 2. In: Dixon, L.C.W., Szegö, G.P. (eds.). North Holland (1978)

Dodge, J., Anderson, C., Smith, N.A.: Random search for hyperparameters using determinantal point processes (2017). arXiv preprint

Engel, Y., Mannor, S., & Meir, R. (2003). Bayes meets Bellman: The Gaussian process approach to temporal difference learning. In Proceedings of the 20th International Conference on Machine Learning (ICML-03) (pp. 154–161)

Evtushenko, Y.G.: Numerical methods for finding global extrema (Case of a non-uniform mesh). USSR Comput. Math. Math. Phys. **11**, 38–54 (1971). https://doi.org/10.1016/0041-5553(71)90065-6

Falkner, S., Klein, A., Hutter, F.: Combining Hyperband and Bayesian Optimization. In: Proceedings of the 31st Conference on Neural Information Processing Systems (NIPS), Bayesian Optimization Workshop (2017)

Florea, A.C., Andonie, R.: A dynamic early stopping criterion for random search in SVM hyper-parameter optimization. In: IFIP International Conference on Artificial Intelligence Applications and Innovations, pp. 168–180. Springer, Cham (2018)

Gershman, S.J.: Deconstructing the human algorithms for exploration. Cognition **173**, 34–42 (2018). https://doi.org/10.1016/j.cognition.2017.12.014

Gershman, S.J.: Uncertainty and exploration. *bioRxiv*, 265504 (2018). https://doi.org/10.1101/265504

Gopnik, A., O'Grady, S., Lucas, C. G., Griffiths, T. L., Wente, A., Bridgers, S., et al.: Changes in cognitive flexibility and hypothesis search across human life history from childhood to adolescence to adulthood. Proc. National Acad. Sci. **114**(30), 7892–7899 (2017)

Griffiths, T.L., Lucas, C., Williams, J., Kalish, M.L. (2009). Modeling human function learning with Gaussian processes. In: Advances in Neural Information Processing Systems, pp. 553–560

Hansen, N.: The CMA evolution strategy: a tutorial (2016). arXiv preprint arXiv:1604.00772

Hauschild, M., Pelikan, M.: An introduction and survey of estimation of distribution algorithms. Swarm Evol. Comput. **1**(3), 111–128 (2011)

Jalali, A., Azimi, J., Fern, X., Zhang, R.: A Lipschitz exploration-exploitation scheme for Bayesian optimization. Lecture Notes in Computer Science (including subseries Lecture Notes in Artificial Intelligence and Lecture Notes in Bioinformatics). 8188 LNAI, 210–224 (2013). https://doi.org/10.1007/978-3-642-40988-2_14

Jones, D.R., Perttunen, C.D., Stuckman, B.E.: Lipschitzian optimization without the Lipschitz constant. J. Optim. Theory Appl. **79**(1), 157–181 (1993)

Li, L., Jamieson, K., DeSalvo, G., Rostamizadeh, A., Talwalkar, A.: Hyperband: a novel bandit-based approach to hyperparameter optimization (2016). arXiv preprint arXiv:1603.06560

Ling, C.K., Low, K.H., Jaillet, P.: Gaussian process planning with Lipschitz continuous reward functions: towards unifying bayesian optimization, active learning, and beyond. In: AAAI, pp 1860–1866 (2016)

Locatelli, M., Schoen, F.: Global Optimization: theory, Algorithms, and Applications, vol. 15. Siam (2013)

Locatelli, M., Schoen, F.: Random linkage: a family of acceptance/rejection algorithms for global optimisation. Math. Program. Series B. **85**, 379–396 (1999). https://doi.org/10.1007/s101070050062

Malherbe, C., Vayatis, N.: Global optimization of Lipschitz functions (2017). arXiv preprint arXiv:1703.02628

Mania, H., Guy, A., Recht, B.: Simple random search provides a competitive approach to reinforcement learning (2018). arXiv preprint arXiv:1803.07055

Missov, T.I., Ermakov, S.M.: On importance sampling in the problem of global optimization. Monte Carlo Methods Appl. **15**, 135–144 (2009). https://doi.org/10.1515/MCMA.2009.007

Norkin, V.I., Pflug, G.C., Ruszczynski, A.: A branch and bound method for stochastic global optimization. Math. Program. **83**, 425–450 (1998)

Ortega, P.A., Wang, J.X., Rowland, M., Genewein, T., Kurth-Nelson, Z., Pascanu, R., et al.: Meta-learning of sequential strategies (2019). arXiv preprint arXiv:1905.03030

Pardalos, P. M., & Romeijn, H. E. (Eds.). (2013). Handbook of global optimization (Vol. 2). Springer Science & Business Media

Parsopoulos, K.E., Vrahatis, M.N.: Particle swarm optimization method in multiobjective problems. In: Proceedings of the 2002 ACM symposium on Applied computing, pp. 603–607. ACM (2002, March)

Pelikan, M., Goldberg, D.E., Cantú-Paz, E.: BOA: the Bayesian optimization algorithm. In: Genetic and Evolutionary Computation Conference, pp. 525–532 (1999)

Peterson, J., Bourgin, D., Reichman, D., Griffiths, T., Russell, S.: Cognitive model priors for predicting human decisions. In International Conference on Machine Learning, pp. 5133–5141 (2019, May)

Pintér, J.D. (ed.).: Global Optimization: scientific and Engineering Case Studies, vol. 85. Springer Science & Business Media (2006)

Powell, W.B., Ryzhov, I.O.: Optimal Learning, vol. 841. Wiley (2012)

Rastrigin, L.A.: The convergence of the random search method in the extremal control of a many parameter system. Autom. Rem. Control. **24**(10), 1337–1342 (1963)

Schoen, F.: Random and quasi-random linkage methods in global optimization. J. Global Optim. **13**, 445–454 (1998). https://doi.org/10.1023/A:1008354314309

Schulz, L.: The origins of inquiry: inductive inference and exploration in early childhood. Trends Cognitive Sciences **16**(7), 382–389 (2012)

Schulz, E., Tenenbaum, J., Duvenaud, D.K., Speekenbrink, M., Gershman, S.J.: Probing the compositionality of intuitive functions. In: Advances in Neural Information Processing Systems, pp. 3729–3737 (2016)

Schulz, E., Tenenbaum, J.B., Reshef, D.N., Speekenbrink, M., Gershman, S.: Assessing the Perceived Predictability of Functions. In: CogSci (2015)

Schulz, E., Speekenbrink, M., Krause, A.: A tutorial on Gaussian process regression: modelling, exploring, and exploiting functions. J. Math. Psychol. **85**, 1–16 (2018a)

Schulz, E., Konstantinidis, E., Speekenbrink, M.: Putting bandits into context: how function learning supports decision making. J. Exp. Psychol.-Learn. Memoru Cognition **44**(6), 927–943 (2018b)

Sergeyev, Y.D., Candelieri, A., Kvasov, D.E., Perego, R.: Safe global optimization of expensive noisy black-box functions in the δ-Lipschitz framework (2019). arXiv preprint arXiv:1908.06010

Sergeyev, Y.D., Kvasov, D.E.: Deterministic Global Optimization: an Introduction to the Diagonal Approach. Springer (2017)

Sergeyev, Y.D., Kvasov, D.E.: Global search based on efficient diagonal partitions and a set of Lipschitz constants. SIAM J. Optim. **16** (2006). https://doi.org/10.1137/040621132

Sergeyev, Y.D., Kvasov, D.E., Mukhametzhanov, M.S.: On the efficiency of nature-inspired metaheuristics in expensive global optimization with limited budget. Scientific Reports **8**(1), 453 (2018)

Shubert, B.O.: A sequential method seeking the global maximum of a function. SIAM J. Numer. Anal. **9**(3), 379–388 (1972)

Srinivas, N., Krause, A., Kakade, S.M., Seeger, M.: Gaussian process optimization in the bandit setting: no regret and experimental design (2009). arXiv preprint arXiv:0912.3995

Strongin, R.G.: Numerical methods in multiextremal problems. Nauka, Moscow, USSR (1978)

Strongin, R.G.: Algorithms for multi-extremal mathematical programming problems employing the set of joint space-filling curves. J. Global Optim. **2**, 357–378 (1992). https://doi.org/10.1007/BF00122428

Torn, A.A.: A search clustering approach to global optimization. Towards Global Optim. **2** (1978)

Ulmer, H., Streichert, F., Zell, A.: Evolution strategies assisted by Gaussian processes with improved preselection criterion. In: The Congress on Evolutionary Computation, CEC'03, vol. 1, pp. 692–699, IEEE (2003, December)

Wang, J., Xu, J., & Wang, X.: Combination of hyperband and Bayesian optimization for hyperparameter optimization in deep learning (2018). arXiv preprint arXiv:1801.01596

Wang, L., Shan, S., Wang, G.G.: Mode-pursuing sampling method for global optimization on expensive black-box functions. Eng. Optim. **36**, 419–438 (2004). https://doi.org/10.1080/03052150410001686486

Wilson, A.G., Dann, C., Lucas, C., Xing, E.P.: The human kernel. In: Advances in Neural Information Processing Systems, pp. 2854–2862 (2015)

Wilson, A., Fern, A., Tadepalli, P.: Using trajectory data to improve bayesian optimization for reinforcement learning. J. Mach. Learn. Res. **15**(1), 253–282 (2014)

Wu, C.M., Schulz, E., Speekenbrink, M., Nelson, J.D., Meder, B.: Generalization guides human exploration in vast decision spaces. Nat. Human Behav. **2**(12), 915 (2018)

Zabinsky, Z.B., Smith, R.L.: Pure adaptive search in global optimization. Math. Program. **53**(1–3), 323–338 (1992)

Zabinsky, Z.B., Wang, W., Prasetio, Y., Ghate, A., Yen, J.W.: Adaptive probabilistic branch and bound for level set approximation. In: Proceedings of the Winter Simulation Conference, pp. 4151–4162. Winter Simulation Conference (2011, December)

Zabinsky, Z.B.: Random search algorithms. In: Wiley Encyclopedia of Operations Research and Management Science (2011)

Zabinsky, Z.B.: Stochastic adaptive search methods: theory and implementation. In: Handbook of Simulation Optimization, pp. 293–318. Springer, New York (2015)

Zhigljavsky, A.: Mathematical theory of global random search. LGU, Leningrad (1985)

Zhigljavsky, A.: Branch and probability bound methods for global optimization. Informatica **1**(1), 125–140 (1990)

Zhigljavsky, A., Zilinskas, A.: Stochastic Global Optimization, vol. 9. Springer Science & Business Media (2007)

Zhigljavsky, A.A., Chekmasov, M.V.: Comparison of independent, stratified and random covering sample schemes in optimization problems. Math. Comput. Model. **23**, 97–110 (1996). https://doi.org/10.1016/0895-7177(96)00043-X

Zielinski, R.: A statistical estimate of the structure of multi-extremal problems. Math. Program. **21**, 348–356 (1981). https://doi.org/10.1007/BF01584254

Žilinskas, A., Zhigljavsky, A.: Stochastic global optimization: a review on the occasion of 25 years of Informatica. Informatica **27**(2), 229–256 (2016)

Žilinskas, A., Gillard, J., Scammell, M., Zhigljavsky, A.: Multistart with early termination of descents. J. Glob. Optim. 1–16 (2019)

Chapter 3
The Surrogate Model

This Chapter presents the first key component of BO, that is, the probabilistic surrogate model. Section 3.1 is focused on Gaussian processes (GPs); Sect. 3.2 introduces the sequential optimization method known as Thompson sampling, also based on GP; finally, Sect. 3.3 presents other probabilistic models which might represent, in some cases, a suitable alternative to GP.

3.1 Gaussian Processes

Gaussian processes are a powerful formalism for implementing both regression and classification algorithms: we focus on regression. While most of the regression algorithms provide a deterministic output, GPs also offer a reliable estimate of uncertainty. This chapter presents the basic mathematics underlying this powerful tool.

3.1.1 Gaussian Processes Regression

One way to interpret a Gaussian process (GP) regression model is to think of it as defining a distribution over functions and with inference taking place directly in the space of functions (i.e. *function-space view*) (Williams and Rasmussen 2006). A GP is a collection of random variables, any finite number of which have a joint Gaussian distribution. A GP is completely specified by its mean function $\mu(x)$ and covariance function $\mathrm{Cov}\big(f(x), f(x')\big) = k(x, x')$:

$$\mu(x) = \mathbb{E}[f(x)]$$
$$\mathrm{Cov}\big(f(x), f(x')\big) = k(x, x') = \mathbb{E}\big[(f(x) - \mu(x))\big(f(x') - \mu(x')\big)\big]$$

and it is defined as:

F. Archetti and A. Candelieri, *Bayesian Optimization and Data Science*, SpringerBriefs in Optimization, https://doi.org/10.1007/978-3-030-24494-1_3

$$f(x) \sim GP(\mu(x), k(x, x'))$$

Usually, for notational simplicity we will take the prior of the mean function to be zero, although this is not necessary.

A simple example of a Gaussian process can be obtained from a Bayesian linear regression model $f(x) = \phi(x)^T w$ with prior $w = \mathcal{N}(0, \Sigma_p)$, where $\phi(x)$ and w are p-dimensional vectors. More precisely, $\phi(x)$ is a function mapping the d-dimensional vector x into a p-dimensional vector.

Thus, the equations for mean and covariance become:

$$\mathbb{E}[f(x)] = \phi(x)^T \mathbb{E}[w] = 0$$
$$\mathbb{E}[f(x)f(x')] = \phi(x)^T \mathbb{E}[ww^T]\phi(x') = \phi(x)^T \Sigma_p \phi(x')$$

This means that $f(x)$ and $f(x')$ are jointly Gaussian with zero mean and covariance given by $\phi(x)^T \Sigma_p \phi(x')$.

As consequence, the function values $f(x_1), \ldots, f(x_n)$ obtained at n different points $x_1, \ldots, x_n$ are jointly Gaussian.

The covariance function assumes a critical role int the GP modelling, as it specifies the distribution over functions. To see this, we can draw samples from the distribution of functions evaluated at any number of points; in detail, we choose a set of input points $X_{1:n} = (x_1, \ldots, x_n)^T$ and then compute the corresponding covariance matrix element wise. This operation is usually performed by using pre-defined covariance functions allowing to write covariance between *outputs* as a function of *inputs* (i.e. $\mathrm{Cov}(f(x), f(x')) = k(x, x')$). Finally, we can generate a random Gaussian vector as:

$$f(X_{1:n}) \sim \mathcal{N}(\mathbf{0}, \mathrm{K}(X_{1:n}, X_{1:n}))$$

and plot the generated values as a function of the inputs. This is basically known as *sampling from prior*, whose core is sampling from a multivariate Gaussian distribution (Tong 1990).

Following an example of five different GP samples drawn from the GP prior: the covariance function used is known as the squared exponential (SE) kernel, introduced in Chap. 1 and that will be detailed in the following Sect. 3.1.2 (Fig. 3.1).

We are usually not primarily interested in drawing random functions from the prior but want to incorporate the knowledge about the function obtained through the evaluations performed so far. Such a knowledge will be then used by the *acquisition function* (presented in Chap. 4) in order to associate an informational utility to each point $x \in X$. We have usually access only to noisy function values, denoted by $y = f(x) + \varepsilon$. Assuming additive independent identically distributed Gaussian noise ε with variance λ^2, the prior on the noisy observations becomes:

$$\mathrm{Cov}(f(x), f(x')) = k(x, x') + \lambda^2 \delta_{xx'}$$

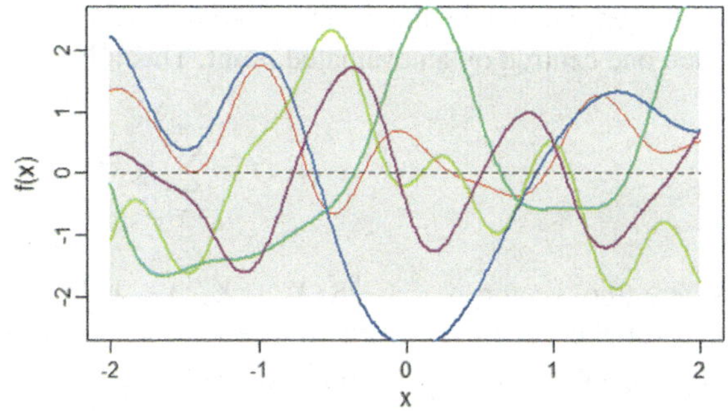

Fig. 3.1 Five different samples from the prior of a GP with squared exponential kernel as covariance function

where $\delta_{xx'}$ is a Kronecker delta which is equal to 1 if and only if $x = x'$. Thus, the covariance over all the function values $\mathrm{y} = (y_1, \ldots, y_n)$ is:

$$\mathrm{Cov}(\mathrm{y}) = \mathrm{K}(X_{1:n}, X_{1:n}) + \lambda^2 I$$

Therefore, the predictive equations for GP regression, that are $\mu(x)$ and $\sigma^2(x)$, can be easily updated, by conditioning the joint Gaussian prior distribution on the observations:

$$\mu(x) = \mathbb{E}[f(x)|D_{1:n}, x] = \mathrm{k}(x, X_{1:n})\big[\mathrm{K}(X_{1:n}, X_{1:n}) + \lambda^2 I\big]^{-1}\mathrm{y}$$
$$\sigma^2(x) = k(x, x) - \mathrm{k}(x, X_{1:n})\big[\mathrm{K}(X_{1:n}, X_{1:n}) + \lambda^2 I\big]^{-1}\mathrm{k}(X_{1:n}, x)$$

These equations are the same reported in Sect. 1.2.1.

Following, a simple example of five different samples drawn at random from a GP prior and posterior, respectively. Posterior is conditioned to six function observations (Fig. 3.2).

Sampling from posterior can be, ideally, considered as generating functions from the prior and rejecting the ones that disagree with the observations. Naturally, this strategy would not be computationally very efficient. A more formal definition of sampling from posterior will be presented in Sect. 3.2.

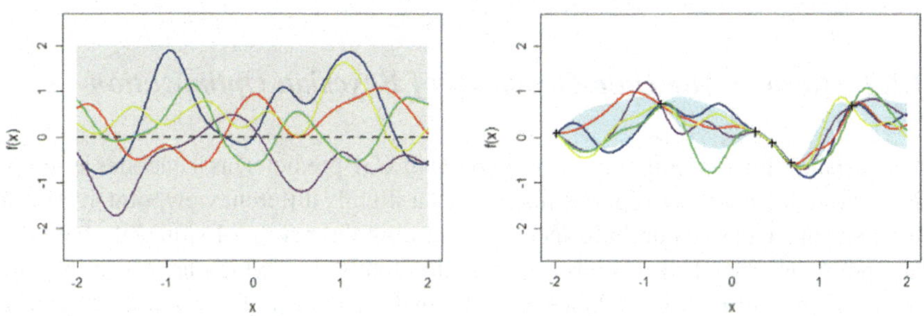

Fig. 3.2 Sampling from prior versus sampling from posterior (for the sake of simplicity, we consider the noise-free setting)

It is easy to notice that the mean prediction is a linear combination of n functions, each one centred on an evaluated point. This allows to write $\mu(x)$ as:

$$\mu(x) = \sum_{i=1}^{n} \alpha_i k(x, x_i)$$

where the vector $\alpha = \left[\mathrm{K}(X_{1:n}, X_{1:n}) + \lambda^2 I\right]^{-1} \mathrm{y}$ and α_i is the i-th component of the vector α, given by the product between the i-th row of the matrix $\left[\mathrm{K}(X_{1:n}, X_{1:n}) + \lambda^2 I\right]^{-1}$ and the vector y.

This means that, to make a prediction at a given x, we only need to consider the $(n+1)$-dimensional distribution defined by the n function evaluations performed so far and the new point x to evaluate.

Several covariance functions have been proposed and some of the most widely adopted will be presented in the following subsection 3.1.2. Every covariance function has some hyperparameters to be set up, defining shape features of the GP, such as smoothness and amplitude. The values of the hyperparameters are usually unknown a priori and are set up depending on the observations $D_{1:n}$. Summarizing with γ the vector of the covariance's hyperparameters, their values are usually set up via marginal likelihood maximization, which is given, in the case of GP for regression, in closed form:

$$p(y \mid X_{1:n}, \gamma) = \int p(y \mid f, X_{1:n}) p(f|X_{1:n}) df$$

The GP's hyperparameters γ appear non-linearly in the kernel matrix K and a closed-form solution maximizing the marginal likelihood cannot be found in general. In practice, gradient-based optimization algorithms are adopted to find a (local) optimum of the marginal likelihood (e.g. conjugate gradients or BFGS).

An interesting generalization has been recently proposed in Berkenkamp et al. (2019) where the values of hyperparameters are modified by iteratively reducing the characteristic length-scale instead of setting them up through marginal likelihood maximization.

3.1.2 *Kernel: The Data Geometry of Bayesian Optimization*

A covariance function is the crucial ingredient in a GP predictor, as it encodes assumptions about the function to approximate. From a slightly different viewpoint, it is clear that both in supervised and unsupervised learning the notion of similarity between data points is crucial; it is a basic assumption that points which are close in x are likely to have similar target values y, and thus function evaluations that are near to a given point should be informative about the prediction at that point. Under the GP view, it is the covariance function that defines nearness or similarity.

A *stationary* covariance function is a function of $x - x'$. Thus, it is invariant to translations in the input space. If further the covariance is a function only of $|x - x'|$ then it is called *isotropic* and it is invariant to all rigid motions. Finally, a *dot product* covariance function depends only on $x \cdot x'$. Dot product covariance functions are invariant to a rotation of the coordinates about the origin but not translations.

A general name for a function k of two arguments mapping a pair of inputs x and x' into a scalar is *kernel*. For a kernel to be a covariance function the following conditions must be satisfied:

- kernel symmetry if $k(x, x') = k(x', x)$
- the matrix K with entries $K_{ij} = k(x_i, x_j)$, also known as *Gram matrix* must be positive semidefinite (PSD).

Examples of covariance (aka kernel) functions:

Squared Exponential (SE) kernel:

$$k_{SE}(x, x') = e^{-\frac{\|x-x'\|^2}{2\ell^2}}$$

with ℓ known as *characteristic length-scale*. The role of this hyperparameter is to rescale any point x by $1/\ell$ before computing the kernel value.

A large length-scale implies long-range correlations, whereas a short length-scale makes function values strongly correlated only if their respective inputs are very close to each other.

This kernel is infinitely differentiable, meaning that the GP is very "smooth". Note, that the covariance between the outputs is written as a function of the inputs. For this particular covariance function, we see that the covariance is almost unity between variables whose corresponding inputs are very close and decreases as their distance in the input space increases.

Matérn kernels:

$$k_{\mathrm{Mat}}(x, x') = \frac{2^{1-\nu}}{\Gamma(\nu)}\left(\frac{|x - x'|\sqrt{2\nu}}{\ell}\right)^{\nu} K_{\nu}\left(\frac{|x - x'|\sqrt{2\nu}}{\ell}\right)$$

with two hyperparameters ν and ℓ, and where K_{ν} is a modified Bessel function. Note that for $\nu \to \infty$ we obtain the SE kernel.

The Matérn covariance functions become, particularly simple when ν is half-integer: $\nu = p + 1/2$, where p is a non-negative integer. In this case, the covariance function is a product of an exponential and a polynomial of order p. The most widely adopted versions, specifically in the machine learning community, are $\nu = 3/2$ and $\nu = 5/2$.

$$k_{\nu=3/2}(x, x') = \left(1 + \frac{|x-x'|\sqrt{3}}{\ell}\right)e^{-\frac{|x-x'|\sqrt{3}}{\ell}}$$

$$k_{\nu=5/2}(x, x') = \left(1 + \frac{|x-x'|\sqrt{5}}{\ell} + \frac{(x-x')^2}{3\ell^2}\right)e^{-\frac{|x-x'|\sqrt{5}}{\ell}}$$

Rational Quadratic Covariance Function:

$$k_{RQ}(x, x') = \left(1 + \frac{(x - x')^2}{2\alpha\ell^2}\right)^{-\alpha}$$

where α and ℓ are two hyperparameters. This kernel can be considered as an infinite sum (*scale mixture*) of SE kernels, with different characteristic length-scales. Indeed, one of the most important properties of kernel functions is that a sum of kernels is a kernel.

The following figure summarizes how the value of the four kernels decreases with x moving away from $x' = 0$ (on the left side) and which are possible resulting samples with different shape properties (on the right side) (Fig. 3.3).

The aforementioned kernels are just the most widely adopted in GP regression. More details and a most comprehensive set of covariance functions are reported in Williams and Rasmussen (2006), including non-stationary kernels and dot product kernels. Kernel for BO is a very actively researched area, relevant papers are Duvenaud et al. (2013; Shilton et al. (2018). Schultz et al. (2016) note that a kernel mismatch about the smoothness of the objective function leads to a substantial drop in accuracy and sample efficiency, which increases with the dimensions and cannot be mitigated by GP's hyperparameter tuning and the choice of acquisition functions.

A space-temporal kernel has been proposed in Nyikosa et al. (2018) to allow the GP to capture all the instances of the function over time and track a temporally evolving minimum. Some kernel issues have been considered in recent publications, such as: kernel composition, safe optimization in relation to cognition (Schultz et al. 2018) as well as kernel learning, adaptation and sparsity in order to deal with functions

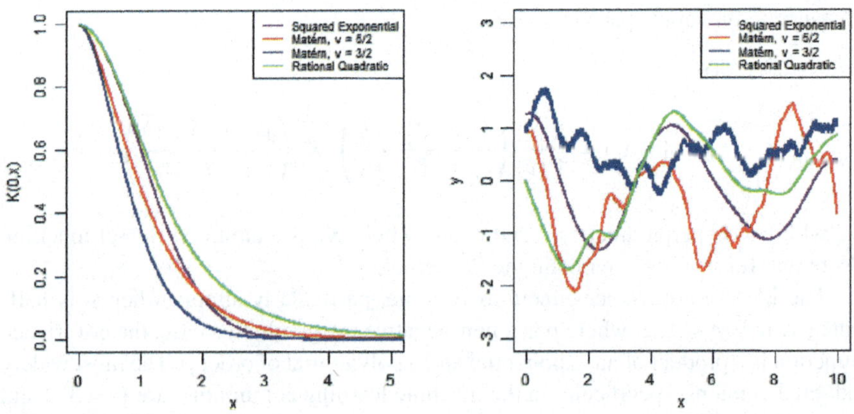

Fig. 3.3 Value of four different kernels with x moving away from $x' = 0$ (left) and four samples from GP prior, one for each kernel considered (right). The value of the characteristic length-scale is ℓ 1 for all the four kernels; α of the RQ kernel is set to 2.25

that are smooth in a subset of their domain and vary rapidly in another is analysed in Peifer et al. (2019).

3.1.3 Embedding Derivative Observations in the Gaussian Process

A relatively new part of GP regression, specifically useful for BO, is the integration of derivative information of $f(x)$ in the modelling and optimization process. Differentiation is a linear operator so that the derivative of a GP is also a GP:, if the covariance function is sufficiently smooth, one can derive the joint distribution over a function and its derivatives and then perform their Bayesian updating given observations of function, derivatives and Hessians (Wu et al. 2017).

We consider a d-dimensional function sampled from a GP, $f(\cdot) \sim \mathcal{GP}(0, K)$. Let us denote the first order partial derivative operator as $\nabla(\cdot) = \left[\frac{\partial}{\partial_{x_1}} \frac{\partial}{\partial_{x_2}} \dots \frac{\partial}{\partial_{x_d}}\right]^T$.

Then the joint process $[f(x), \nabla f(x)]$ has a GP distribution (see for instance Solak et al. (2003)) with a covariance function given by four blocks:

$$\begin{aligned} k_{[f,f]}(x, x') &= \text{cov}(f(x), f(x')) = k(x, x') \\ k_{[f,\nabla f]}(x, x') &= \text{cov}(f(x), \nabla f(x')) = \nabla_{x'} k(x, x') \\ k_{[\nabla f, f]}(x, x') &= \text{cov}(\nabla f(x), f(x')) = \nabla_x k(x, x') \\ k_{[\nabla f,\nabla f]}(x, x') &= \text{cov}(\nabla f(x), \nabla f(x')) = \nabla_x \nabla_{x'} k(x, x') \end{aligned}$$

We can write this joint process more compactly as

$$\begin{bmatrix} f(x) \\ \nabla f(x) \end{bmatrix} \sim \mathcal{GP}\left(0, \begin{bmatrix} \mathrm{k} & \mathrm{k}_{[f,\nabla f]} \\ \mathrm{k}_{[\nabla f, f]} & \mathrm{k}_{[\nabla f,\nabla f]} \end{bmatrix}\right)$$

We can now apply the formulas to update $\mu(x)$ and $\sigma(x)$ and to derive the posterior over function values $f(x_{n+1})$ given a set of observations of function values and gradients. Let $\mathrm{K}_{[\mathrm{f},\nabla \mathrm{f}]}$ denote the joint kernel matrix for a set of observations of function values and gradients. The joint distribution for $\left[f(X_{1:n}), \nabla f(X_{1:n}), f(x_{n+1})\right]$ is

$$\begin{bmatrix} f(X_{1:n}) \\ \nabla f(X_{1:n}) \\ f(x_{n+1}) \end{bmatrix} \sim \mathcal{N}\left(0, \begin{bmatrix} \mathrm{K}_{[\mathbf{f},\nabla \mathbf{f}]} & \bar{\mathrm{k}}_{n+1} \\ \bar{\mathrm{k}}_{n+1}{}^{\mathsf{T}} & k(x_{n+1}, x_{n+1}) \end{bmatrix}\right)$$

where $\bar{\mathrm{k}}_{n+1} = \left[\mathrm{k}(x_{n+1}, X_{1:n})^{\mathsf{T}}, \mathrm{k}_{[f_{n+1},\nabla f]}\right]^{\mathsf{T}}$, and the posterior over $f(x_{n+1})$ is:

$$f(x_{n+1}) | x_{1:n}, [f, \nabla f]_{1:n}, x_{n+1} \sim \mathcal{N}\left(\bar{\mu}(x_{n+1}), \bar{\sigma}^2(x_{n+1})\right)$$

where

$$\bar{\mu}(x_{n+1}) = \bar{\mathrm{k}}_{n+1}^{\mathsf{T}} \mathrm{K}_{[f,\nabla f]}^{-1} [f, \nabla f]_{1:n}^{\mathsf{T}}$$
$$\bar{\sigma}^2(x_{n+1}) = k(x_{n+1}, x_{n+1}) - \bar{\mathrm{k}}_{n+1}^{\mathsf{T}} \mathrm{K}_{[f,\nabla f]}^{-1} \bar{\mathrm{k}}_{n+1}$$

We use a similar derivation to incorporate Hessian (second derivative) information into a GP along with the function and gradient information.

Then the joint Gaussian of $\left[f, \nabla f, \nabla^2 f(x)\right]$ is defined as,

$$\begin{bmatrix} f(x) \\ \nabla f(x) \\ \nabla^2 f(x) \end{bmatrix} \sim \mathcal{GP}\Big(0, \mathrm{K}_{[f,\nabla f,\nabla^2 f]}\Big)$$

where

$$\mathrm{K}_{[f,\nabla f,\nabla^2 f]} = \begin{bmatrix} \mathrm{k} & \mathrm{k}_{[f,\nabla f]} & \mathrm{k}_{[f,\nabla^2 f]} \\ \mathrm{k}_{[\nabla f,f]} & \mathrm{k}_{[\nabla f,\nabla f]} & \mathrm{k}_{[\nabla f,\nabla^2 f]} \\ \mathrm{k}_{[\nabla^2 f,f]} & \mathrm{k}_{[\nabla^2 f,\nabla f]} & \mathrm{k}_{[\nabla^2 f,\nabla^2 f]} \end{bmatrix}$$

The resulting joint kernel matrix is partitioned into nine blocks corresponding to the covariances and cross-covariances over function values, gradients and Hessians.

We can now derive the posterior over function values $f(x_{n+1})$ given a set of observations of function, its gradients and Hessians. The joint distribution of $\left[f(X_{1:n}), \nabla f(X_{1:n}), \nabla^2 f(X_{1:n}), f(x_{n+1})\right]$ is given by:

$$\begin{bmatrix} f(X_{1:n}) \\ \nabla f(X_{1:n}) \\ \nabla^2 f(X_{1:n}) \\ f(x_{n+1}) \end{bmatrix} \sim \mathcal{N}\left(0, \begin{bmatrix} \mathrm{K}_{[f,\nabla f,\nabla^2 f]} & \bar{\mathrm{k}}_{n+1} \\ \bar{\mathrm{k}}_{n+1}{}^{\mathsf{T}} & k(x_{n+1}, x_{n+1}) \end{bmatrix}\right)$$

where $\bar{\mathrm{k}}_{n+1} = \left[\mathrm{k}(x_{n+1}, X_{1:n})^{\mathsf{T}}, \mathrm{k}_{[f_{n+1},\nabla f]}, \mathrm{k}_{[f_{n+1},\nabla^2 f]}\right]^{\mathsf{T}}$ and the posterior over $f(x_{n+1})$ is:

$$f(x_{n+1}) | x_{1:n}, \left[f, \nabla f, \nabla^2 f\right]_{1:n}, x_{n+1} \sim \mathcal{N}\big(\bar{\mu}(x_{n+1}), \bar{\sigma}^2(x_{n+1})\big)$$

where

$$\bar{\mu}(x_{n+1}) = \bar{\mathrm{k}}_{n+1}^{\mathsf{T}} \mathrm{K}_{[f,\nabla f,\nabla^2 f]}^{-1} \left[f, \nabla f, \nabla^2 f\right]_{1:n}^{\mathsf{T}}$$
$$\bar{\sigma}^2(x_{n+1}) = k(x_{n+1}, x_{n+1}) - \bar{\mathrm{k}}_{n+1}^{\mathsf{T}} \mathrm{K}_{[f,\nabla f,\nabla^2 f]}^{-1} \bar{\mathrm{k}}_{n+1}$$

The idea of using derivative information for optimization problems has been taken up in Hennig and Kiefel (2013; Hennig (2013) and in Wills and Thomas (2017) which has developed it along two different directions. The first, originally suggested in the papers by Henning, brings to the reinterpretation in probabilistic terms of standard quasi-Newton like methods considered as particular instances of Gaussian process or Bayesian regression. The second, more relevant for the subject of this book, is to use the joint GP as a global model of the objective function and its derivatives.

A first remark is that adding a derivative observation reduces the uncertainty-standard deviation of GP prediction. The following figure, drawn from Solak et al. (2003) shows, from left to right, four samples from prior, four samples from posterior of standard GP and four samples from posterior of a GP with derivative observations. Given the same set of observations, the variance of the resulting GP is lower when derivative observations are included (Fig. 3.4).

The same pay-off effect of gradient estimation is shown in the following picture, drawn from Eriksson et al. (2018). On the left, the level sets of the objective function (Branin 2D) is depicted, in the middle and on the right, the mean of the GP without and with derivative observations, is reported, respectively (Fig. 3.5).

The same authors propose a BO algorithm with derivatives, inspired by REMBO (Wang et al. 2016). The algorithm estimates the active subspace spanned by the dominant eigenvalues of the gradient covariance matrix and fits a GP on the set of observations in this subspace. The optimization of the acquisition function,

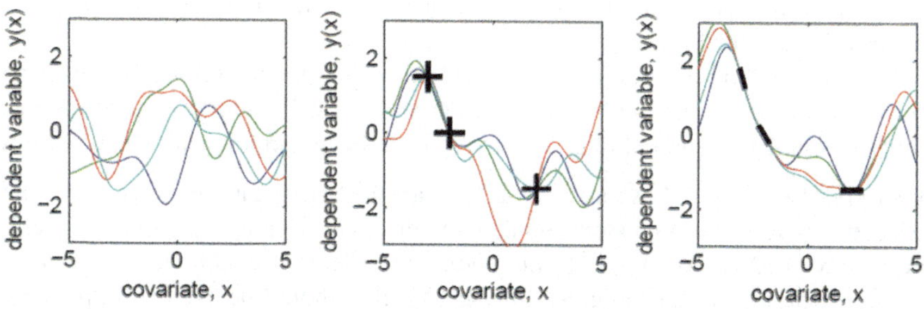

Fig. 3.4 Four samples from prior, four samples from posterior of a standard GP and four samples from posterior of a GP with derivative observations. (*Source* Solak et al. 2003)

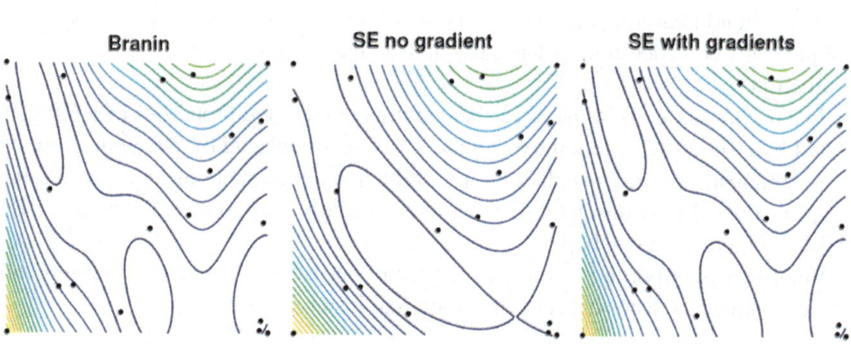

Fig. 3.5 Original function (left) mean of a GP without derivative observations (middle) and mean of a GP with derivative observations. (*Source* Eriksson et al. 2018)

built on the updated model, brings the new point x_{n+1}. The dataset is updated with $\left[x_{n+1}, f(x_{n+1}), \nabla f(x_{n+1})\right]$ and, consequently, the GP's hyperparameters are updated considering also the information on gradients.

3.1.4 Numerical Instability

Numerical instability issues arise due to the inversion of the matrix $\left[\mathrm{K}(X_{1:n}, X_{1:n}) + \lambda^2 I\right]$, required to update $\mu(x)$ and $\sigma^2(x)$. The *condition number* of this matrix, that is the ratio between the highest and lowest eigenvalue, gives a bound on the accuracy of the matrix inversion. Large condition numbers are indicators for numerical instability. Extreme eigenvalues appear when function values are strongly correlated. Strong positive correlation between function values will result in near-identical corresponding rows and columns in the kernel matrix $\mathrm{K}(X_{1:n}, X_{1:n})$, making it close to singular. Higher values of λ, on one side mitigate this instability, on the other side lead to a degradation of the model's accuracy.

Some heuristics have been proposed to control the condition number of the matrix $\left[\mathrm{K}(X_{1:n}, X_{1:n}) + \lambda^2 I\right]$. The most common consists of adding an explicit penalty on the signal-to-noise ratio, when fitting the GP or adding a jitter term. Both are presented in https://drafts.distill.pub/gp/.

The first method derives from the observation that the condition number grows linearly with the number n of data points and quadratically with the signal-to-noise ratio $\frac{\sigma_f}{\lambda}$ (where σ_f^2 is a multiplier used to regulate the maximum kernel value). More precisely, $\frac{e_{max}}{e_{min}} = n\frac{\sigma_f^2}{\lambda^2}+1$, where e_{max} and e_{min} are the largest and smallest eigen value, respectively. It is easy to understand that a high signal-to-noise ratio makes quickly the matrix $\left[\mathrm{K}(X_{1:n}, X_{1:n}) + \lambda^2 I\right]$ ill-conditioned. Therefore, although most of the research works consider the noise-free setting, this should be avoided in practical applications for numerical stability reasons. When fitting the GP by maximizing the log-marginal likelihood, a penalty term is added to the log-marginal likelihood that discourages extreme signal-to-noise ratios.

The second heuristic method, consisting of adding a jitter term, is also based on the previous consideration: it basically adds some artificial "noise" to the function evaluations.

In Zhigljavsky and Žilinskas (2019), authors studied the properties of a GP generated bya SE kernel where the exponent 2 has been replaced by 2-ε, with ε > 0.

An interesting solution, called K-optimality and aimed at choosing the next point x_{n+1} to reduce the condition number, has been recently suggested in Yan et al. (2018) and will be outlined in Chap. 4.

Finally, it is important to remark that using derivatives, as reported in previous section, induces a faster onset of the numerical instability issue.

3.2 Thompson Sampling

As previously mentioned, given a kernel k, there exists a feature map $\phi(x)$ such that $k(x, x') = \phi(x)^T\phi(x')$. In practice, one can use the spectral decomposition of the matrix $K = \mathbf{\Phi}\mathbf{\Phi}^T$ to estimate the feature map $\phi(x) \in \mathbb{R}^m$, where $\mathbf{\Phi}^T = [\phi(x_1), \ldots, \phi(x_n)]$ and $\{x_1, \ldots, x_n\}$ are the points evaluated so far. The decomposition we considered is derived from the Bochner's theorem, as reported in (Basu and Ghosh 2017), and based on the concept of *spectral density* of a kernel. For instance, the spectral density *S(s)* of the SE covariance function is given by:

$$S(s) = (2\pi\ell^2)^{\frac{d}{2}} e^{-2\pi^2\ell^2 s^2}$$

The spectral density can be treated as a probability density $p(s) = S(s)/\alpha$ with $\alpha = \int S(s)ds$ a normalizing constant. The following figure shows *p(s)* for different values of the value of the length scale of the SE kernel (Fig. 3.6).

In TS, the feature map *φ(x)* is a m-dimensional vector sampled as $\sqrt{2/m}\cos(\mathbf{W}x + \mathbf{b})$, where $[\mathbf{W}]_i \sim p(s)$ and $[\mathbf{b}]_i \sim \mathcal{U}(0, 2\pi)$, with i denoting the ith component of the vectors $\mathbf{W}$ and $\mathbf{b}$, and $i = 1, \ldots, m$.

Accordingly, the following figures show how the shape of the (prior) sample modifies depending on the length scale (i.e. ℓ is 0.2, 0.3 and 0.4, from left to right). Every sample is generated by sampling $[\mathbf{W}_i] \sim p(s)$ and $[\mathbf{b}]_i \sim \mathcal{U}(0, 2\pi)$ (Figs. 3.7 and 3.8).

Although Thompson sampling (TS) dates to (Thompson 1933), it has been recently attracting attention in several studies on sequential optimization (Chapelle and Li 2011; Kandasamy et al. 2017, 2018). Ouyang et al. (2017) propose a TS-based

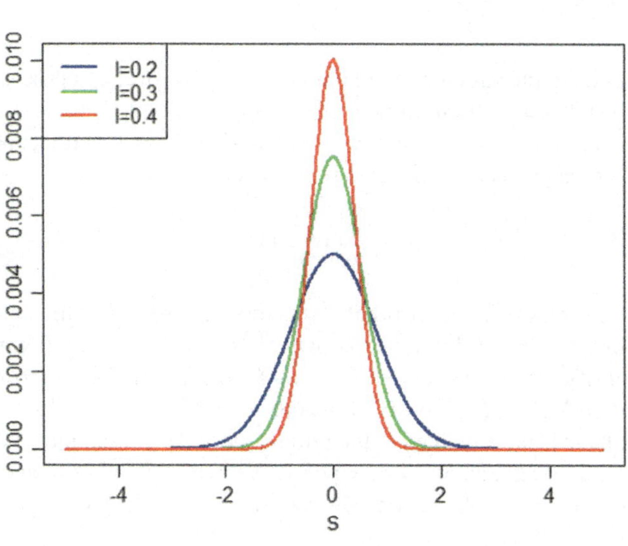

Fig. 3.6 Spectral density of SE kernel with different values of the characteristic length-scale

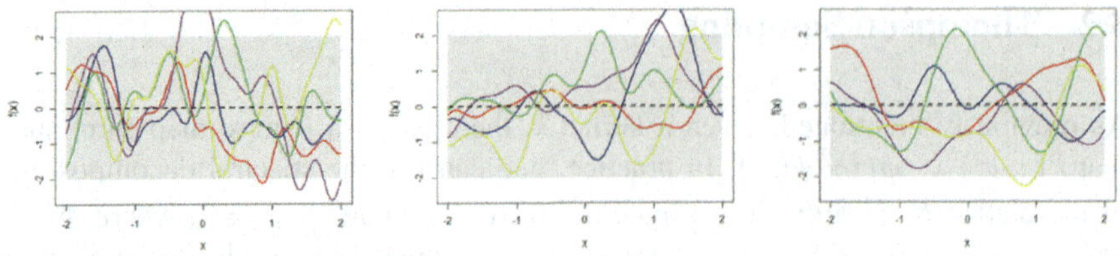

Fig. 3.7 Five different samples for GP prior for, respectively, $\ell = 0.2$, $\ell = 0.3$ and $\ell = 0.4$, respectively

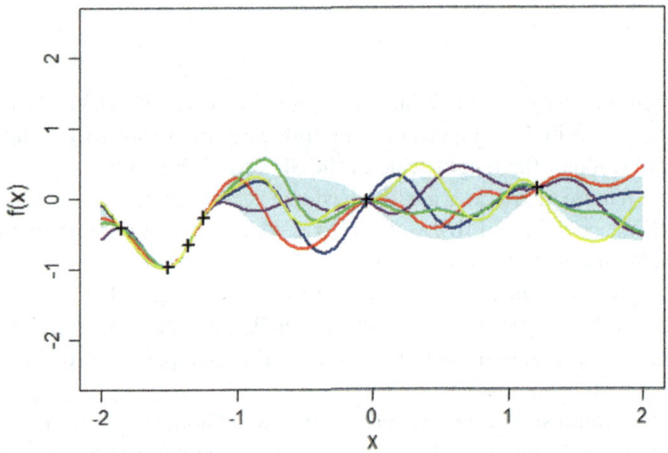

Fig. 3.8 Five different samples for GP posterior (ℓ is learned through marginal likelihood maximization on the observations)

approach to learn the structure of an unknown Markov decision process (MDP) in a reinforcement learning framework.

TS is an algorithm to sequentially optimize a black box function, coherently with the optimization problem considered in this book:

$$\min_{x \in X \subset \mathbb{R}^d} f(x)$$

As BO, TS uses a GP as a probabilistic surrogate model of the unknown objective function. The main difference is that TS samples, at every iteration, just one mapping function $\phi(x)$ and a random vector $\boldsymbol{\theta}$ drawn from the posterior distribution $\boldsymbol{\theta}|(D_{1:n}, \phi) = \mathcal{N}\left(\boldsymbol{A}^{-1}\boldsymbol{\Phi}^T \boldsymbol{y}, \sigma^2 \boldsymbol{A}^{-1}\right)$, where $\boldsymbol{A} = \boldsymbol{\Phi}^T\boldsymbol{\Phi} + \sigma^2 \boldsymbol{I}$. Then the new point x_{n+1} to evaluate is just obtained as the minimizer of the GP sample $f(x) = \phi(x)^T \boldsymbol{\theta}$. The function is evaluated at the new point, eventually with noise, and the function evaluations dataset is updated consequently: $D_{n+1} = D_n \cup \{(x_{n+1}, y_{n+1})\}$.

The process is initialized by assuming a non-informative prior on the distribution of the minimizer and will stop when the variance of the distribution becomes considerably small.

The following figure shows four different samples from GP posterior (after five observations) and the corresponding minimizers to use as next point to evaluate. We have decided to plot four different scenarios to highlight the probabilistic nature of TS in selecting the next point. Although samples are different, giving some chance to exploration, the bias towards exploitation is evident, motivating the ε-greedy approach proposed in (Basu and Ghosh 2017) (Fig. 3.9).

As a final remark, it is clear that TS is a sequential optimization process per se, and sampling from posterior is just one component of the method. It is again based on a probabilistic surrogate model but, differently from BO, the next promising point to evaluate is identified by minimizing a sample from the GP instead of a specific acquisition function. Another way to look at this is to consider the sample function as an acquisition function (as described in Chap. 4).

While the theory underlying TS and related converge issue are rooted in results in functional analysis, like Bochner's theorem and Winer–Khintchine theorem, its implementation is relatively straightforward as shown before and demonstrated in the pseudo code which follows, again inspired by (Basu and Ghosh 2017). There are different implementations, notable (Bijl et al. 2016)

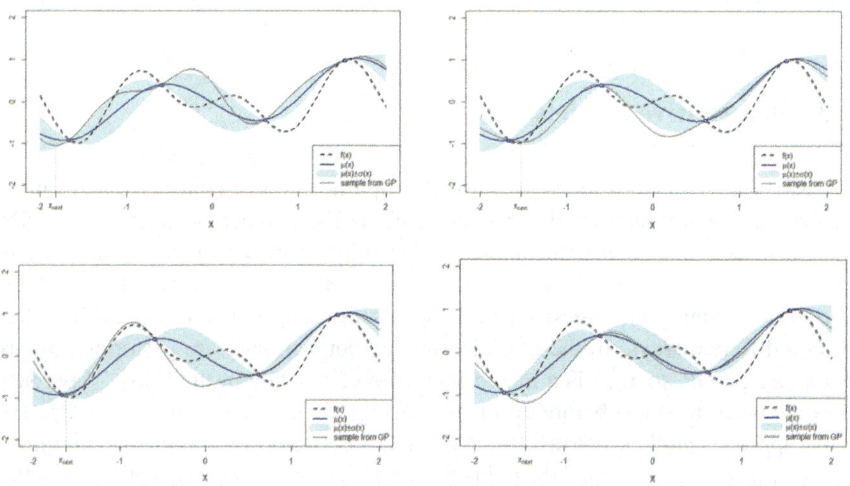

Fig. 3.9 An example of one iteration of Thompson sampling: four different samples for GP posterior are reported along with the next point to evaluate chosen as the minimizer of the corresponding GP sample

Algorithm - Thompson Sampling with SE kernel

Input: function $f(x)$, kernel k, domain X
Output: x^+ the optimal solution minimizing $f(x)$
Observe $y_1 \leftarrow f(x_1) + \varepsilon$, with $\varepsilon \sim \mathcal{N}(0, \lambda^2)$
$D_1 \leftarrow \{(x_1, y_1)\}$
$x^+ \leftarrow x_1$
$y^+ \leftarrow f(x^+)$
for $n = 1, 2, \ldots$ **do**
  estimate the kernel's hyperparameters depending on $D_{1:n}$(e.g., the length scale ℓ in the case of SE kernel)
  sample a random feature map $\phi(x)$ (i.e., from $\sqrt{2/m}\cos(\mathbf{W}\mathbf{x} + \mathbf{b})$, as previously described)
  sample $\boldsymbol{\theta}$ from $\boldsymbol{\theta}|(D_{1:n}, \phi) \sim \mathcal{N}(\boldsymbol{A}^{-1}\boldsymbol{\Phi}^T\boldsymbol{y}, \sigma^2\boldsymbol{A}^{-1})$, with $\boldsymbol{A} = \boldsymbol{\Phi}^T\boldsymbol{\Phi} + \sigma^2\boldsymbol{I}$ and $\boldsymbol{\Phi}^T = [\phi(x_1), \ldots, \phi(x_n)]$
  select $x_{n+1} = \text{argmin}_{x \in X}\, \phi(x)^T\boldsymbol{\theta}$
  observe $y_{n+1} \leftarrow f(x_{n+1}) + \varepsilon$, with $\varepsilon \sim \mathcal{N}(0, \lambda^2)$
  $D_{n+1} \leftarrow D_n \cup \{(x_{n+1}, y_{n+1})\}$
  break from the loop when x_{n+1} converges to x^+
  if $y_{n+1} < y^+$ **then**
    $x^+ \leftarrow x_{n+1}$
    $y^+ \leftarrow y_{n+1}$
  end
endfor
return x^+

TS can be biased towards exploitation, especially when σ^2 is large, leading the process to potentially converge to a local minimum. To reduce the risk to get stuck in a local optimum, an ε-greedy approach can be considered, as in (Basu and Ghosh 2017). That is, at every iteration n, we explore the entire region X uniformly (i.e. we sample $x_{n+1} \sim \mathcal{U}(X)$), with probability $\varepsilon > 0$ or we sample $x_{n+1} \sim min_{x \in X}\phi(x)^T\boldsymbol{\theta}$, with probability $1 - \varepsilon$.

3.3 Alternative Models

A basic issue with GP is the assumption that optimization variables take real values. In the case of integer-valued variables, such as the number of layers of a DNN, the basic idea is rounding the solution to the closest integer while, in the case of categorical variables, as many extra variables as possible categories are added and chosen according to the largest one (as in Spearmint). These operations can be applied in two different ways. In the so-called naïve approach they are performed before updating the GP, so x_{n+1} is a mixed-integer vector, while in the "basic" approach they are performed just before evaluating the objective function, so $x_{n+1} \in \mathbb{R}^d$. The naïve approach might lead to a relevant mismatch between the (continuous) optimizer of the objective function and the point evaluated. The basic approach overcomes this limitation because GP ignores the approximation but it is expected to provide a suboptimal solution. A more principled approach is suggested in Garrido-Merchán and Hernández-Lobato (2018), who proposes a kernel that essentially embeds the same transformation of the basic approach.

A radical solution consists in avoiding GP altogether and adopting alternative models like random forests.

3.3.1 Random Forest

Random forest (RF) is an ensemble learning method, based on decision trees, for both classification and regression problems (Ho 1995). According to the originally proposed implementation, RF aims at generating a multitude of decision trees, at training time, and providing as output the mode of the classes (classification) the mean/median prediction (regression) of the individual trees. The following figure provides a schematic representation of the output provisioning mechanism of a RF (Fig. 3.10).

Decision trees are ideal candidates for ensemble methods since they usually have low bias and high variance, making them very likely to benefit from the averaging process. RF mostly differs from other typical ensemble methods in the way it introduces random perturbations into the training phase. The basic idea is to combine bagging, to sample examples from the original dataset, and random selection of features, in order to train every tree of the forest. Injecting randomness simultaneously with both strategies yields one the most effective off-the-shelf methods in machine learning, working surprisingly well for almost any kind of problems, allowing for generating a collection of decision trees with controlled variance.

Although designed and presented as a machine learning algorithm, RF is also an effective and efficient alternative to GP for implementing BO. To better understand how RF, for regression, can replace GP, one has to consider that:

- The dataset, in the case of Machine Learning, is replaced by the design $D_{1:n}$ of the BO process
- The features correspond to the dimensions of the global optimization problem.

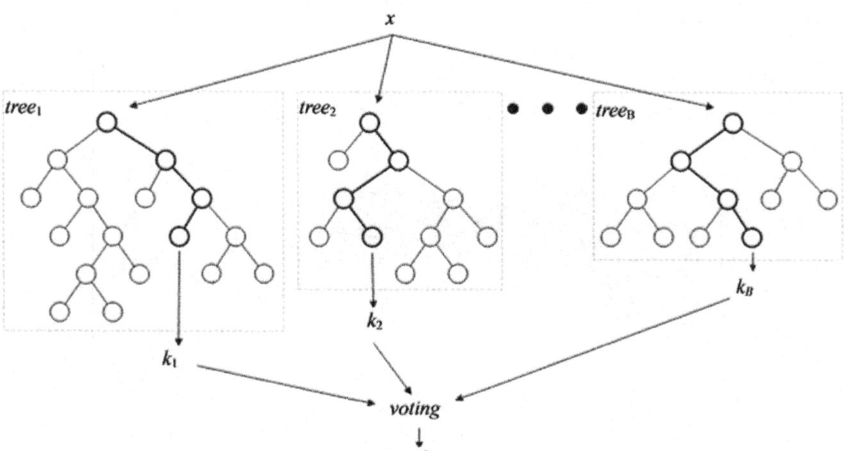

Fig. 3.10 A schematic representation of the regression/classification performed by a random forest. Voting mechanism is different for classification (e.g. simple or weighted majority) and regression (e.g. mean or median)

Moreover, since RF consists of an ensemble of different regressors, it is possible to compute—as for GP—both $\mu(x)$ and $\sigma(x)$, simply as mean and variance among the individual outputs provided by the regressor. Due to the different nature of RF and GP, the associated probabilistic surrogate models will also result significantly different. The following figure offers an example. While GP is well-suited to model smooth functions in search space spanned by continuous variables, RF can deal with discrete and conditional variables. Discontinuity in the RF-based surrogate is given by the underlying decision tree models (Fig. 3.11).

Another important property of RF is that the variance can be controlled, as explained in the following.

Let's denote with S the size of the forest (i.e. the number of decision trees in the forest) and $T_i(x)$ the output provided by the i-th decision tree for the input x, the variance of the RF-based estimator, for the regression case, can be easily computed as follows:

$$
\begin{aligned}
\mathrm{Var}\left(\frac{1}{S}\sum_{i=1}^{S} T_i(x)\right) &= \mathrm{Cov}\left(\frac{1}{S}\sum_{i=1}^{S} T_i(x), \frac{1}{S}\sum_{j=1}^{S} T_j(x)\right) \\
&= \frac{1}{S^2}\sum_{i=1}^{S}\left(\sum_{j\neq i}^{S} \mathrm{Cov}(T_i(x), T_j(x)) + \mathrm{Var}(T_i(x))\right) \\
&\leq \frac{1}{S^2}\sum_{i=1}^{S}\left((S-1)\rho\sigma^2 + \sigma^2\right) \\
&= \frac{S(S-1)\rho\sigma^2 + S\sigma^2}{S^2} = \rho\sigma^2 + \sigma^2\frac{1-\rho}{S}
\end{aligned}
$$

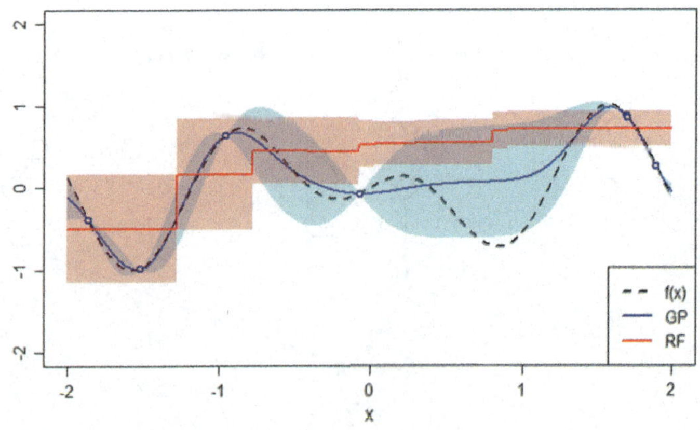

Fig. 3.11 A graphical comparison between probabilistic surrogate models offered by a GP (blue) and a RF (red), both fitted on the same set of observations

where σ^2 is the maximum variance computed over all the $T_i(x)$ and $\rho\sigma^2 = \max_{i,j} \mathrm{Cov}\big(T_i(x), T_j(x)\big)$. The variance of the RF estimator is proportional to σ^2 and ρ (i.e. if the number m of selected features decreases also σ^2 and ρ decrease) and with the size S the forest increasing.

Finally, a basic description of the RF learning algorithm is provided. Thanks to the bagging and random feature selection—step 2 and 4, respectively—RF results more computationally efficient than GP, specifically when the number of decision variables is larger than 20. Indeed, RF does not require to invert any kernel matrix and training of every decision tree can be performed in parallel.

Random Forest learning algorithm
Given:
· S the size of the forest (i.e. number of decision trees in the forest)
· N the number of examples in the dataset (i.e. evaluations of the objective functions)
· M the number of features (i.e. decision variables) representing every example
· m the number of features (i.e. decision variables) to be randomly selected from the M available for each leaf node of the tree to be split
· $n_{\min}$ the minimum node size in the tree
1: for $i = 1$ to S
2:   sample, with replacement, a subset of N instances from the dataset
3:   for every terminal node (leaf) of the i-th decision tree with size less than $n_{\min}$
4:     select m features (i.e. decision variables) at random from the M available
5:     pick the best feature (i.e. decision variable) and split among the possible m
6:     split the node in two children nodes (new leaves of the decision tree)
7:   end for
8: end for

3.3.2 *Neural Networks: Feedforward, Deep and Bayesian*

The role of neural networks in BO is two-fold: as already discussed they are machine learning algorithms typically optimized in the AutoML setting; on the other hand, they can offer a well-suited alternative to GP. Indeed, as already highlighted in this book, GPs are sample efficient but their computational complexity is cubic in the number of points and do not scale well in high dimensions.

A relevant approach, Snoek et al. (2015) aims to replace the GP with a model that scales better but retains most of the GP desirable properties such as flexibility and well-calibrated uncertainty. More specifically, deep learning models are investigated adding a Bayesian linear regressor to the last hidden layer of a deep neural network (DNN), marginalizing only the output weights of the net while using a point estimate for the remaining parameters. This results in adaptive basis regression, a well-established statistical technique which scales linearly in the number of observations, and cubically in the basis function dimensionality. This allows to explicitly trade-off evaluation time and model capacity. The resulting algorithm DNGO (Deep Networks for Global Optimization, https://github.com/Anmol6/DNGO-BO) has been extensively tested for optimizing the hyperparameters of deep convolutional neural networks. Empirical results show that DNGO provides the same modelling properties of a GP but with a significantly lower computational cost.

The following figure (Fig. 3.12) shows ANN and DNN for creating alternative surrogate models within BO framework and allows for a comparison of predictive mean (solid blue line) and uncertainty (shaded blue region). The first line figures show the surrogate models generated by DNN with three hidden layers, the first one using only tanh as activation function and the second one using only sigmoid as the activation function. The second line figures show the surrogate models generated by ANN with one hidden layer with the same activation functions

The application of Bayesian methods to neural networks has a rich history in machine learning, the goal of Bayesian neural networks is to uncover the full posterior distribution over the network weights in order to capture uncertainty, to act as a regularizer, and to provide a framework for model comparison. The full posterior

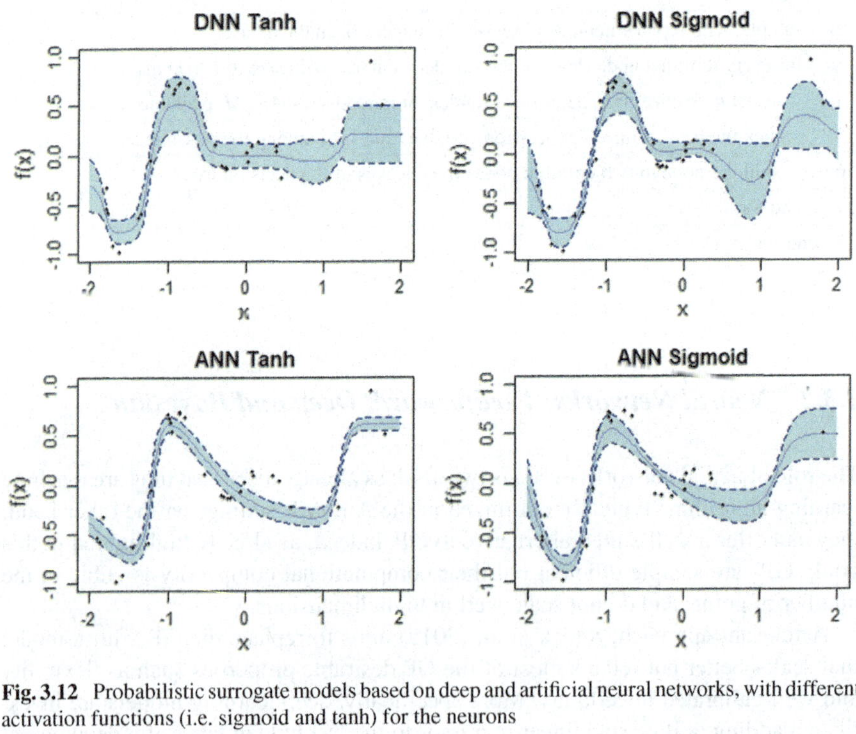

Fig. 3.12 Probabilistic surrogate models based on deep and artificial neural networks, with different activation functions (i.e. sigmoid and tanh) for the neurons

is, however, intractable for most forms of neural networks, necessitating expensive approximate inference or Markov Chain Monte Carlo simulation.

In Springenberg et al. (2016), BNN have been suggested as a principled alternative to GP. The algorithm is called BOHAMIANN (Bayesian Optimization with HAMIltonian Artificial Neural Network) and has been tested with a three layers neural network with 50 tanh units. The method has been presented also for multi-task optimization (i.e. finding the set of optimizers for k black box functions, each with the same domain X). An implementation of the algorithm is provided in RoBO, a BO software reported in Chap. 6.

In order to deal with non-stationarity, a possible approach is to use deep Gaussian processes. In Hebbal et al. (2019), functional composition of stationary GPs is proposed, providing a multiple layer structure.

References

Basu, K., Ghosh, S.: Analysis of Thompson sampling for Gaussian process optimization in the bandit setting (2017). arXiv preprint arXiv:1705.06808

Bergstra, J., Bardenet, R., Bengio, Y., Kégl, B.: Algorithms for hyper-parameter optimization. In: Advances in Neural Information Processing Systems, pp. 2546–2554 (2011)

Berkenkamp, F., Schoellig, A.P., Krause, A.: No-regret Bayesian Optimization with unknown Hyperparameters(2019). arXiv preprint arXiv:1901.03357

Bijl, H., Schön, T.B., van Wingerden, J.W., Verhaegen, M.: A sequential Monte Carlo approach to Thompson sampling for Bayesian optimization (2016). arXiv preprint arXiv:1604.00169

Chapelle, O., Li, L.: An empirical evaluation of Thompson sampling. In: NIPS, pp. 2249–2257 (2011)

Duvenaud, D., Lloyd, J.R., Grosse, R., Tenenbaum, J.B., Ghahramani, Z.: Structure discovery in nonparametric regression through compositional kernel search (2013). arXiv preprint arXiv:1302.4922

Eriksson, D., Dong, K., Lee, E., Bindel, D., Wilson, A.G.: Scaling Gaussian process regression with derivatives. In: Advances in Neural Information Processing Systems, pp. 6868–6878 (2018)

Garrido-Merchán, E.C., Hernández-Lobato, D.: Dealing with categorical and integer-valued variables in Bayesian optimization with Gaussian processes (2018). arXiv preprint arXiv:1805.03463

Hebbal, A., Brevault, L., Balesdent, M., Talbi, E.G., Melab, N.: Bayesian Optimization using Deep Gaussian Processes (2019). arXiv preprint arXiv:1905.03350

Hennig, P., Kiefel, M.: Quasi-Newton method: a new direction. J. Mach. Learn. Res. **14**, 843–865 (2013)

Hennig, P.: Fast probabilistic optimization from noisy gradients. In: International Conference on Machine Learning, pp. 62–70 (2013, February)

Ho, T.K.: Random decision forests. In: Conference in Document Analysis and Recognition, pp. 278–282 (1995)

Kandasamy, K., Krishnamurthy, A., Schneider, J., Póczos, B. Parallelised bayesian optimisation via Thompson sampling. In: International Conference on Artificial Intelligence and Statistics, pp. 133–142 (2018, March)

Kandasamy, K., Krishnamurthy, A., Schneider, J., Poczos, B.: Asynchronous parallel Bayesian optimisation via Thompson sampling (2017). arXiv preprint arXiv:1705.09236

Nyikosa, F.M., Osborne, M.A., Roberts, S. J.: Bayesian optimization for dynamic problems (2018). arXiv preprint arXiv:1803.03432

Ouyang, Y., Gagrani, M., Nayyar, A., Jain, R.: Learning unknown markov decision processes: a Thompson sampling approach. In: Advances in Neural Information Processing Systems, pp. 1333–1342 (2017)

Peifer, M., Chamon, L., Paternain, S., Ribeiro, A.: Sparse multiresolution representations with adaptive kernels (2019). arXiv preprint arXiv:1905.02797

Shilton, A., Gupta, S., Rana, S., Vellanki, P., Li, C., Park, L., Venkatesh, S., Sutti, A., Rubin, D., Dorin, T., Vahid, A., Height, M.:. Covariance function pre-training with m-kernels for accelerated Bayesian optimisation (2018). arXiv preprint arXiv:1802.05370

Schulz, E., Speekenbrink, M., Krause, A.: A tutorial on Gaussian process regression: modelling, exploring, and exploiting functions. J. Math. Psychol. **85**, 1–16 (2018)

Snoek, J., Rippel, O., Swersky, K., Kiros, R., Satish, N., Sundaram, N., Patwary, M., Prabhat, M., Adams, R.: Scalable bayesian optimization using deep neural networks. In: International Conference on Machine Learning, pp. 2171–2180 (2015, June)

Schulz, E., Speekenbrink, M., Hernández-Lobato, J.M., Ghahramani, Z., Gershman, S.J.: Quantifying mismatch in Bayesian optimization. In: Nips Workshop on Bayesian Optimization: black-box Optimization and Beyond (2016)

Solak, E., Murray-Smith, R., Leithead, W.E., Leith, D.J., Rasmussen, C.E.: Derivative observations in Gaussian process models of dynamic systems. In: Advances in Neural Information Processing Systems, pp. 1057–1064 (2003)

Springenberg, J.T., Klein, A., Falkner, S., Hutter, F.: Bayesian optimization with robust Bayesian neural networks. In: Advances in Neural Information Processing Systems, pp. 4134–4142 (2016)

Tong Y.L.: Fundamental properties and sampling distributions of the multivariate normal distribution. In: The Multivariate Normal Distribution. Springer Series in Statistics. Springer, New York, NY (1990)

Thompson, W.R.: On the likelihood that one unknown probability exceeds another in view of the evidence of two samples. Biometrika **25**, 285–294 (1933)

Wang, Z., Hutter, F., Zoghi, M., Matheson, D., de Feitas, N.: Bayesian optimization in a billion dimensions via random embeddings. J. Artif. Intell. Res. **55**, 361–387 (2016)

Williams, C.K., Rasmussen, C.E.: Gaussian Processes for Machine Learning, vol. 2, no. 3, p. 4. MIT Press, Cambridge (2006)

Wills, A.G., Schön, T.B.: On the construction of probabilistic Newton-type algorithms. In: IEEE 56th Annual Conference on Decision and Control (CDC), pp. 6499–6504, IEEE. (2017, December)

Wu, A., Aoi, M.C., Pillow, J.W.: Exploiting gradients and Hessians in Bayesian optimization and Bayesian quadrature (2017). arXiv preprint arXiv:1704.00060

Yan, L., Duan, X., Liu, B., Xu, J.: Bayesian optimization based on k-optimality. Entropy **20**(8), 594 (2018)

Zhigljavsky, A., Žilinskas, A.: Selection of a covariance function for a Gaussian random field aimed for modeling global optimization problems. Optim. Lett. 1–11 (2019)

Chapter 4
The Acquisition Function

The acquisition function is the mechanism to implement the trade-off between *exploration* and *exploitation* in BO. More precisely, any acquisition function aims to guide the search of the optimum towards points with potential low values of objective function either because the prediction of $f(x)$, based on the probabilistic surrogate model, is low or the uncertainty, also based on the same model, is high (or both). Indeed, *exploiting* means to consider the area providing more chance to improve the current solution (with respect to the current surrogate model), while *exploring* means to move towards less explored regions of the search space where predictions based on the surrogate model are more uncertain, with higher variance.

This chapter is devoted to present some of the most relevant acquisition functions, from the "traditional" towards the most recent ones.

4.1 Traditional Acquisition Functions

This section summarizes the most widely used acquisition functions.

4.1.1 *Probability of Improvement*

Probability of improvement (PI) was the first acquisition function proposed in the literature Kushner (1964):

$$PI(x) = P\left(f(x) \leq f\left(x^+\right)\right) = \Phi\left(\frac{f\left(x^+\right) - \mu(x)}{\sigma(x)}\right)$$

F. Archetti and A. Candelieri, *Bayesian Optimization and Data Science*, SpringerBriefs in Optimization, https://doi.org/10.1007/978-3-030-24494-1_4

where $f(x^+)$ is the best value of the objective function observed so far, $\mu(x)$ and $\sigma(x)$ are mean and standard deviation of the probabilistic surrogate model, such as a GP, and $\Phi(\cdot)$ is the normal cumulative distribution function.

One of the drawbacks of PI is that it is biased towards exploitation. To mitigate this effect, one can introduce the parameter ξ which modulates the balance between exploration and exploitation. The resulting equation is:

$$\mathrm{PI}(x) = P(f(x) \leq f(x^+) + \xi) = \Phi\left(\frac{f(x^+) - \mu(x) - \xi}{\sigma(x)}\right)$$

Low values of ξ favour exploitation while large values exploration.

Finally, the next point to evaluate is chosen according to: $x_{n+1} = \mathrm{argmax}_{x \in X}\, \mathrm{PI}(x)$.

However, a weak point of PI is to assign a value to a new point irrespective of the potential magnitude of the improvement. This is the reason why the next acquisition function was proposed.

4.1.2 Expected Improvement

Expected improvement (EI) was proposed initially in Mockus et al. (1978) and then made popular in Jones et al. (1998) which measures the expectation of the improvement on $f(x)$ with respect to the predictive distribution of the probabilistic surrogate model.

$$\mathrm{EI}(x) = \begin{cases} (f(x^+) - \mu(x))\Phi(Z) + \sigma(x)\phi(Z) \text{ if } \sigma(x) > 0 \\ 0 \text{ if } \sigma(x) = 0 \end{cases}$$

where $\phi(Z)$ and $\Phi(Z)$ are the probability distribution and the cumulative distribution of the standardized normal, respectively, where

$$Z = \begin{cases} \frac{f(x^+) - \mu(x)}{\sigma(x)} \text{ if } \sigma(x) > 0 \\ 0 \text{ if } \sigma(x) = 0 \end{cases}$$

The EI is made up of two terms: the first is increased by decreasing the predictive mean; the second by increasing the predictive uncertainty. Thus, EI, in a sense, automatically balances, respectively, exploitation and exploration. When we want to actively manage the trade-off between exploration and exploitation, we can introduce the parameter ξ. When exploring, points associated with high uncertainty of the probabilistic surrogate model are more likely to be chosen, while when exploiting, points associated with low value of the mean of the probabilistic surrogate model are selected.

$$\mathrm{EI}(x) = \begin{cases} \left(f\left(x^{+}\right) - \mu(x) - \xi\right)\Phi(Z) + \sigma(x)\phi(Z)\ \mathrm{if}\ \sigma(x) > 0 \\ 0\ \mathrm{if}\ \sigma(x) = 0 \end{cases}$$

and

$$Z = \begin{cases} \frac{f\left(x^{+}\right) - \mu(x) - \xi}{\sigma(x)}\ \mathrm{if}\ \sigma(x) > 0 \\ 0\ \mathrm{if}\ \sigma(x) = 0 \end{cases}$$

The following figure shows how the selected points change depending on ξ. Ideally, ξ should be adjusted dynamically to decrease monotonically with the function evaluations. Compared to PI (Fig. 4.1), it is possible to notice that EI (Fig. 4.2) is less biased towards exploitation.

Finally, the next point to evaluate is chosen according to: $x_{n+1} = \mathrm{argmax}_{x \in X}\ \mathrm{EI}(x)$.

EI has been largely used since 1998 and specialized to specific contexts: Astudillo and Frazier (2019) propose EI-CF a version of composite functions which leads to a multi-output GP: the authors also note that constrained optimization can be regarded as a special case of the optimization of composite functions and that EI-CF reduces

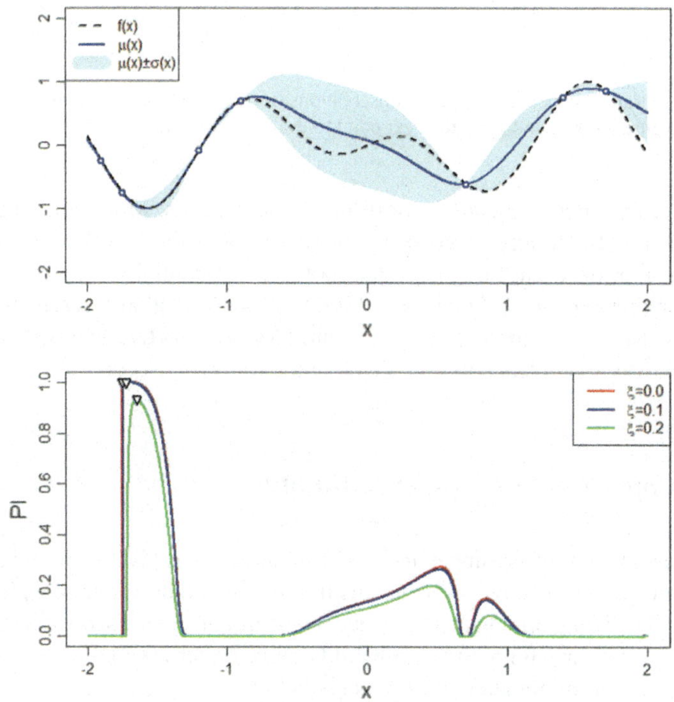

Fig. 4.1 GP trained depending on seven observations (top), PI with respect to different values of ξ and max values corresponding to the next point to evaluate (bottom)

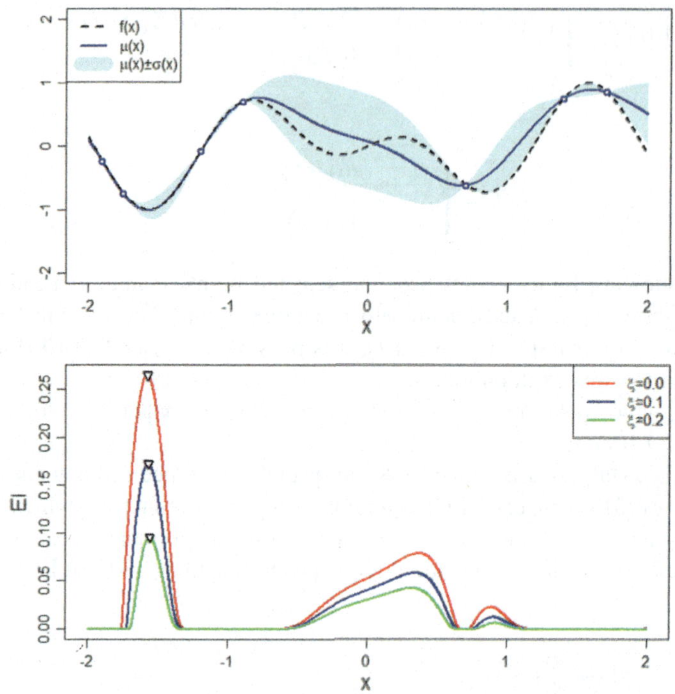

Fig. 4.2 GP trained depending on seven observations (top), EI with respect to different values of ξ and max values corresponding to the next point to evaluate (bottom)

to the expected improvement for constrained optimization. More importantly, both PI and EI are structurally exploitative, increasing ξ from 0 to 0.2 - which is the 10% of the min-max range of $f(x)$ - does not shift substantially x_{n+1}. Higher values of ξ flatten the acquisition function making it close to Random Search. To manage effectively the exploitation-exploration balance a more effective acquisition function is Upper-Lower Confidence Bound, given in the following section.

4.1.3 Upper/Lower Confidence Bound

Confidence bound—where upper and lower are used, respectively, for maximization and minimization problems—is an acquisition function that manages exploration—exploitation by being optimistic in the face of uncertainty, in the sense of considering the best-case scenario for a given probability value (Auer 2002).

For the case of minimization, LCB is given by:

$$\mathrm{LCB}(x) = \mu(x) - \xi\sigma(x)$$

where $\xi \geq 0$ is the parameter to manage the trade-off between exploration and exploitation ($\xi = 0$ is for pure exploitation; on the contrary, higher values of ξ emphasizes exploration by inflating the model uncertainty). For this acquisition function, there are strong theoretical results, originated in the context of multi-armed bandit problems, on achieving the optimal regret derived by Srinivas et al. (2012). For the candidate point x_n, we observe instantaneous regret $r_n = f(x_n) - f(x^*)$. The cumulative regret R_N after N function evaluations is the sum of instantaneous regrets: $R_N = \sum_{n=1}^{N} r_n$. A desirable asymptotic property of an algorithm is to be no-regret: $\lim_{N\to\infty} \frac{R_N}{N} = 0$. Bounds on the average regret $\frac{R_N}{N}$ translate to convergence rates: $f\left(x^+\right) = \min x_{n \leq N} f(x_n)$ in the first N function evaluations is no further from $f(x^*)$ than the average regret. Therefore, $f\left(x^+\right) - f(x^*) \to 0$, with $N \to \infty$.

The following figure shows how the selected points changes depending on ξ. Ideally, ξ should be adjusted dynamically to decrease monotonically with the function evaluations. Compared to PI (Fig. 4.1) and EI (Fig. 4.2), is clear the leaning of LCB towards exploration and how this effect can be managed through the value of ξ (Fig. 4.3).

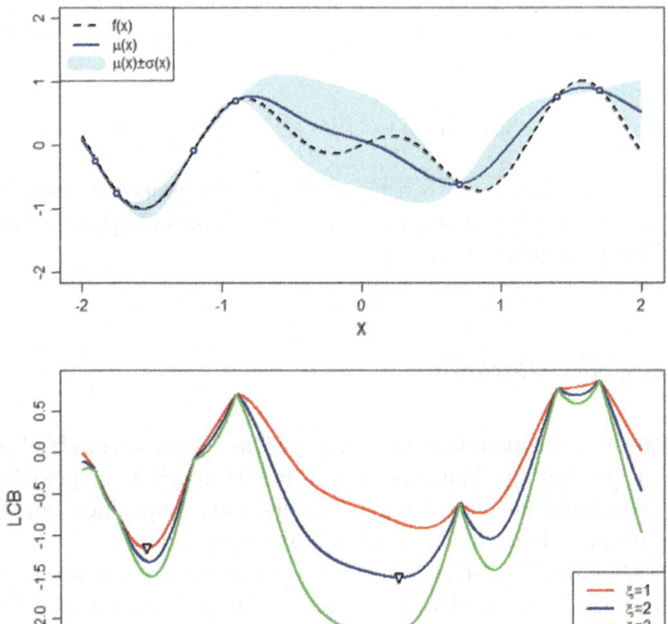

Fig. 4.3 GP trained depending on seven observations (top), LCB with respect to different values of ξ and min values corresponding to the next point to evaluate (bottom). Contrary to the other acquisition functions, LCB is minimized instead of maximized

Finally, the next point to evaluate is chosen according to $x_{n+1} = \text{argmin}_{x \in X} \text{LCB}(x)$, in the case of a minimization problem, or $x_{n+1} = \text{argmax}_{x \in X} \text{UCB}(x)$ in the case of a maximization problem.

4.2 New Acquisition Functions

4.2.1 *Scaled Expected Improvement*

Recently, a new improvement-based acquisition function has been proposed in Noè & Husmeier (2018). The motivation is that EI does not account for the uncertainty in improvement. The proposed acquisition function chooses the next point to evaluate where the improvement is expected to be high with small variance.

More precisely, the variance of the improvement is computed analytically as:

$$\mathbb{V}[I(x)] = k(x,x)\left[\left(Z^2+1\right)\Phi(Z) + Z\phi(Z)\right] - [EI(x)]^2$$

where Z is defined as in EI. The new acquisition function consists in scaling EI by $\{\mathbb{V}[I(x)]\}^{1/2}$, that is:

$$\text{Scaled EI}(x) = \text{EI}(x)/\{\mathbb{V}[I(x)]\}^{1/2}$$

The main result on a set of benchmarks is that the proposed acquisition function compares favourably with PI, EI, UCB/LCB and the entropy-based acquisition functions to be presented in Sect. 4.2.4.

4.2.2 *Portfolio Allocation*

Several acquisition functions have been proposed, as reported in this book, so choosing a good acquisition function is not trivial. Indeed, there are no guidelines in the choice of acquisition function. Furthermore, the same acquisition function is not necessarily the best choice on the entire optimization process.

An interesting mixed strategy has been proposed where the acquisition function is chosen, adaptively at each iteration, from a "portfolio" (Brochu et al. 2010). The selection mechanism is formalized as an on online multi-armed bandit problem: different strategies are investigated with the result that a hierarchical hedging approach is the winning one. The basic algorithm, so-called GP-Hedge, is as follows:

$D_{1:n} = \{(x_1, y_1), \ldots, (x_n, y_n)\}$ is the initial set of evaluated point

N is the overall number of available function evaluations

Set initial "gains" $g_n = 0$ with $g_n \in \mathbb{R}^M$, where M is the number of acquisition functions in the portfolio

for $i = n + 1, \ldots, N$ do

  compute the next point to evaluate, according to each acquisition function: $x_{i,k} = \underset{x}{\operatorname{argmax}}\, \alpha_k(x|D_{1:n}), k = 1, \ldots, M$

  select one point x_i, among the alternative $x_{i,k}$, with probability: $p_{i,k} = \frac{e^{g_{i[k]}}}{\sum_{l=1}^{M} e^{g_{i[l]}}}$ (i.e., soft – max policy) where $g_{i[k]}$ is the gain of the kth acquisition function at iteration i

  evaluate objective function, eventually with noise, $y_i = f(x_i) + \varepsilon$

  update the function evaluations dataset $D_{1:i} = D_{1:n} \cup \{(x_i, y_i)\}$

  update GP and update gains $g_{i[k]} = g_{i-1[k]} + \mu_i(x_{i,k})$

endfor

Empirical evidence shows a significant improvement over EI and LCB, which is anyway offset by the much higher computational cost.

4.2.3 *Thompson Sampling*

As described in Chap. 3, Thompson Sampling (TS) is a sequential optimization process based on performing, iteratively, the following block of steps: updating a posterior depending on a set of observations, drawing a sample from posterior as an approximation to the function to be optimized, minimizing this sample function to identify the next candidate point and evaluating the objective function at that point.

In a sense TS is in itself a sequential decision-making process, like BO, indeed the acquisition function is replaced by a sample from the current probabilistic surrogate model. However, in the literature, TS is considered just as an acquisition function in BO—this is the reason why we have decided to reserve a small section also in this chapter.

In the following figure, the difference in choosing the next point, depending on TS and LCB, is depicted. The (TS) sample function from the GP is quite different from LCB and, at least in this case, TS is less explorative than LCB. Indeed, TS is in principle exploitative due to the sample minimization step. The exploration is given by its Monte Carlo basis and can be enhanced by adopting an ε-greedy strategy (as already seen in Chap. 3) (Fig. 4.4).

The following figure shows another case where TS is, instead, more explorative than LCB (Fig. 4.5).

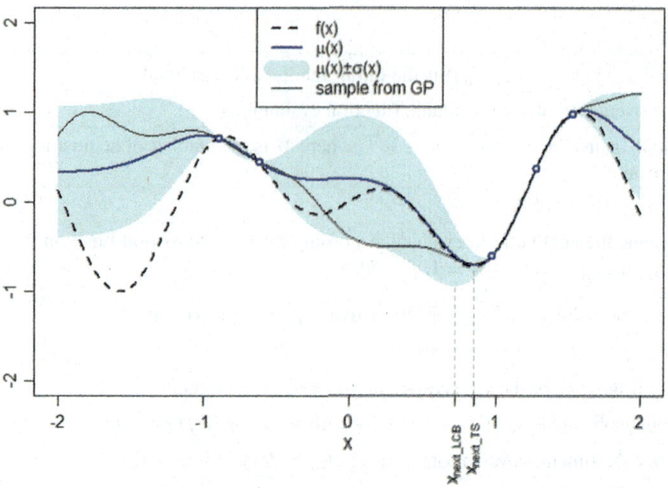

Fig. 4.4 Next point to evaluate according to TS ($\varepsilon = 0.4$) and LCB: the sample from GP posterior implies a more exploitative choice than LCB

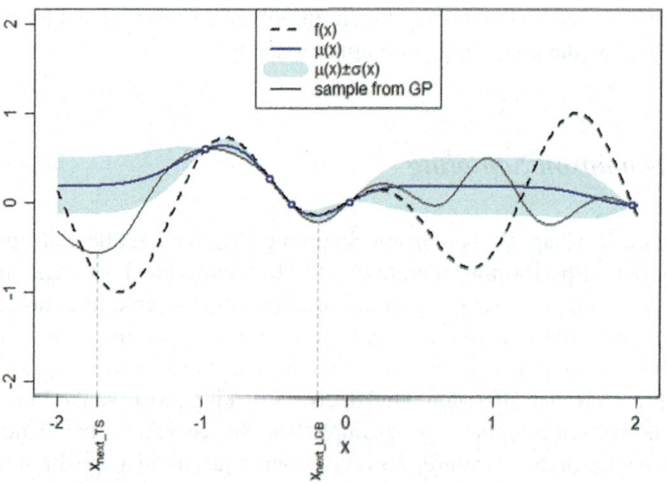

Fig. 4.5 Next point to evaluate according to TS ($\varepsilon = 0.4$) and LCB: the sample from GP posterior implies a more explorative choice than LCB

Theoretical analysis was provided for the classical multi-armed bandit problem and later extended to the analysis of the continuous case in Russo et al. (2018) drawing on an analogy between TS and UCB.

4.2.4 Entropy-Based Acquisition Functions

The traditional acquisition functions, presented in Sect. 4.1, are based on probabilistic measures of improvement in the f domain. Exploitation and exploration are represented, respectively, by the mean value and the "uncertainty bonus" represented by the variance. Recent interest has been focused on querying at points that can help to learn most about the location of the unknown minimum, leading to information-based acquisition functions. This informational approach was originally proposed in Villemonteix et al. (2009), and developer into the Entropy Search (Henning and Schuler (2012)), Predictive Entropy Search (Hernández-Lobato et al. 2014). In both Entropy Search (ES) and Predictive Entropy Search (PES), the basic idea is to maximize the information about the global optimizer.

The current information about the global optimizer can be computed as the negative differential entropy of $p(x^*|D_{1:n})$ and the next point to evaluate is given by the maximization of the expected reduction in this quantity. The ES and PES acquisition functions have the following equations, respectively:

$$\text{ES}: \alpha(x) = H\big(p(x^*|D_{1:n})\big) - \mathbb{E}\big[H\big(p(x^*|D_{1:n} \cup \{x, y\})\big)\big]$$

$$\text{PES}: \alpha(x) = H(p(y|D_{1:n}, x)) - \mathbb{E}\big[H\big(p(y|D_{1:n}, x, x^*)\big)\big]$$

where $H[p(\alpha)] = -\int p(\alpha)\log p(\alpha)d\alpha$ represents the differential entropy.

While ES uses the expectation is over $p(x^*|D_{1:n})$, in PES the expectation is over $p(y|D_{1:n}, x)$. Both $p(x^*|D_{1:n})$ and its entropy are analytically intractable and must be approximated through expensive simulation. This approach needs several approximations and is anyway beset by numerical difficulties. To add further complexity to the computation of this distribution, even if the global optimizer is unique, it is not stable meaning that a small perturbation can result in an approximation far away from the global optimum. PES is inspired by Thompson sampling and uses several samples from posterior, at every iteration, to solve probabilistically the optimization of the acquisition function.

The following figure depicts how PES improves over ES, when compared to the ground truth (Fig. 4.6).

In Wang and Jegelka (2017), a computationally more effective solution is provided, based on sampling from the conditional distributions of a global optimum given the observed data. A much cheaper and more robust acquisition function, namely Max-value Entropy Search (MES) is proposed: instead of measuring the information about the optimizer x^*, they use information about the optimal value y^*. MES measures the gain in mutual information between the optimal value y^* and

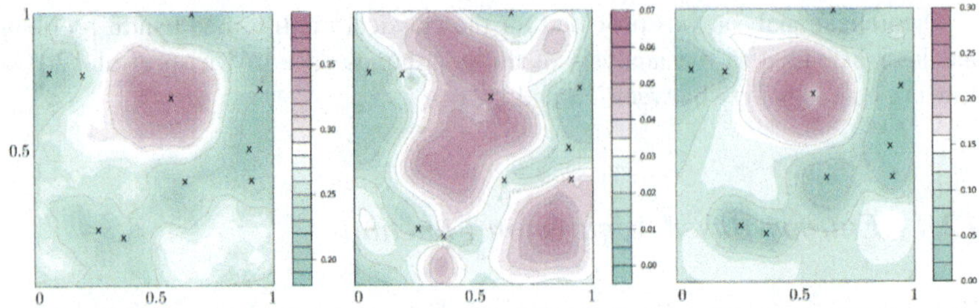

Fig. 4.6 Ground truth (left), approximation produced by the ES method (middle) and approximation produced by the PES method (right). (*Source* Hernandez-Lobato et al. 2014)

the next point to evaluate, which can be approximated analytically by evaluating the entropy of the predictive distribution:

$$\text{MES} : \alpha(x) = H(p(y|D_{1:n}, x)) - \mathbb{E}\left[H\left(p\left(y|D_{1:n}, x, y^*\right)\right)\right]$$

In MES, the expectation is approximated using Monte Carlo estimation. Importantly, while ES and PES rely on the expensive d-dimensional distribution $p(x^*|D_{1:n})$, MES relies on the one-dimensional $p(y^*|D_{1:n})$ which is computationally much easier. The software for PES and MES can be downloaded at https://bitbucket.org/jmh233/codepesnips2014 and https://github.com/zi-w/Max-value-Entropy-Search, respectively. In the case that the value of f^* is known a priori, a more efficient search for x^* is propsed in Nguyen and Osborne (2019).

According to another approach Volpp et al. (2019), the acquisition function is given by a neural network which uses as input the posterior prediction of the GP.

4.2.5 Knowledge Gradient

Knowledge gradient (KG) is an acquisition function based on revising the typical assumption made in the BO process that the best evaluated point is returned as final solution. KG revises this assumption by allowing to return any point as a solution, even if it has not been previously evaluated. Another important difference with other acquisition functions, which are usually "risk-seeker", is that KG assumes risk neutrality (Berger (2013), meaning that any random x is evaluated depending on the expected value $f(x)$.

Let us denote this solution with $\bar{x}_n$; the value $f(\bar{x}_n)$ is random under the posterior and has the following expected value conditioned to $D_{1:n}$:

$$\mu_n^* := \mu_n(\bar{x}_n) = \min_{x'} \mu_n(x')$$

If a further function evaluation were available, we could sample x one more time and obtain the additional observation (x_{n+1}, y_{n+1}) and, consequently, the updated posterior mean $\mu_{n+1}(\cdot)$. The expected value of the solution after this further function evaluation would be $\mu_{n+1}^* = \min_{x'} \mu_{n+1}(x')$. Thus, the improvement in conditional expected solution value is given by: $\mu_n^* - \mu_{n+1}^*$. As this quantity is unknown, before evaluating $f(x_{n+1})$, we can only compute its expected value, conditioned to the previous observations at $x_1, \ldots, x_n$. This quantity is named knowledge gradient and can be computed at any x:

$$KG_n(x) = \mathbb{E}[\mu_n^* - \mu_{n+1}^* | x_{n+1} = x]$$

The simplest way to compute KG is through simulation: a possible value y_{n+1} is simulated at a certain x_{n+1}, then the optimum of the new posterior mean μ_{n+1}^* is computed, by considering the sampled value y_{n+1} as the actual value obtained through function evaluation at x_{n+1}. The simulation can be performed either by Thompson sampling or directly by using $\mu_n(x_{n+1})$ and $\sigma_n(x_{n+1})$. Finally, this value is subtracted from μ_n^* to obtain the corresponding improvement in solution quality. The entire procedure is repeated many times and the differences $\mu_n^* - \mu_{n+1}^*$ are averaged on the simulated values y_{n+1}. This allows to compute the expected value required for the computation of the estimate of $\text{KG}_n(x)$, that converges to the actual value as the number of samples increases.

Therefore, contrary to other acquisition functions, KG computes the posterior not only at the sampled points but on the entire domain, estimating how a new function evaluation will change the posterior. KG assigns a positive value on measurements that cause the minimum of the posterior mean to improve. Thus, the next point to evaluate is given by the maximization of KG.

According to Frazier et al. (2009), this provides a small performance benefit in the case of BO with noise-free evaluations, but a substantial improvement in problems with noisy, derivative, multi-fidelity observations and other "exotic" problem features. In these cases, the value of sampling comes not through an improvement in the best solution at the sampled point, but through an improvement in the minimum of the posterior mean across possible solutions. For instance, a derivative observation can provide information that the function is decreasing, along with a specific direction, in the neighbourhood of the sampled point. Consequently, the minimum of the posterior mean could be significantly smaller than the previous minimum, even if the function value at the sampled point is worse than the best previously sampled point. In these cases, KG can reportedly outperform EI (Wu et al. 2017; Poloczek et al. 2017; Wu and Frazier 2016; Toscano-Palmerin and Frazier 2018) (Fig. 4.7).

The same variance reduction effect that gradients availability has on GP—as shown in Chap. 3—is also observed in acquisition function. In particular, the following figure shows that both KG and EI, computed on a GP with derivatives, make different sampling decisions (Wu and Frazier 2017) (Fig. 4.8).

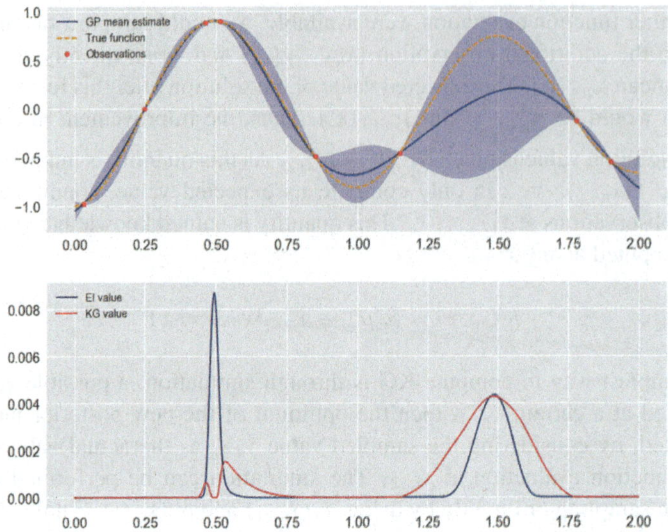

Fig. 4.7 Comparison between KG and EI on a maximization problem. The EI acquisition function prefers to sample in the region around 0.5 and the KG policy prefers to sample in the region around 1.5. Thus, KG gives a chance to exploration while EI is biased towards exploitation. (*Source* https://sigopt.com/blog/expected-improvement-vs-knowledge-gradient/)

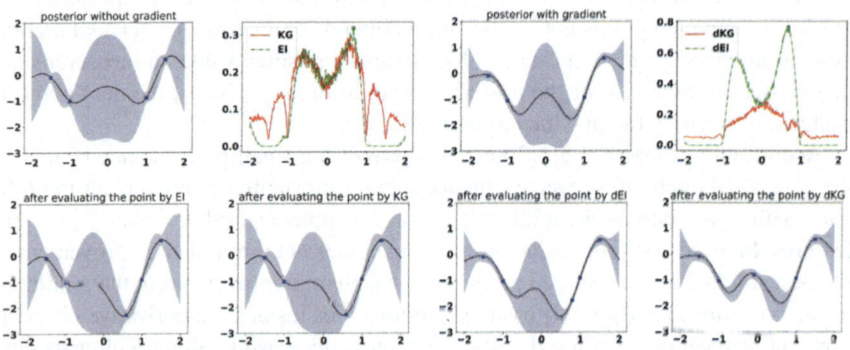

Fig. 4.8 Plots in the top row show the approximations of $f(x)$ modelled through a GP without and with incorporating gradients observations. The posterior variance is smaller if the gradients are incorporated. Both KG and EI are depicted, with and without embedding derivative information. The plots in the bottom row depict the resulting GPs, without and with gradient, after the evaluation at the new point suggested by EI and KG. (*Source* Wu and Frazier 2017)

4.2.6 *Look-Ahead*

Most of the acquisition functions consider only the impact of the next function evaluation: in this sense, they are also called "myopic". This is a limitation addressed in several papers: in Osborne et al. (2009), a two-step-ahead solution is presented; while in Marchant et al. (2014), partially observable Markov decision process (POMDP) is used as a model and a Monte Carlo tree search is adopted to solve it. Another look-ahead acquisition function has been proposed in (Lam et al. 2016) where approximate dynamic programming is adopted to solve the problem of selecting the next candidate point to evaluate. More specifically, a rollout heuristic was proposed for BO.

Another approach, GLASSES, has been suggested in Gonzalez et al. (2016), proposing to model the n-step look-ahead problem as a Bayesian network and then suggesting a computationally efficient approximation to solve the otherwise prohibitively computational expensive inference process. At our knowledge, GLASSES is no longer maintained, last update is from November 2015, based on Python 2.7, while other tools presented in Chap. 6 are now based on Python 3.0 and higher.

4.2.7 K-Optimality

A very interesting approach has been recently proposed in Yan et al. (2018), addressing the issue of numerical instability in the kernel matrix K. If the next point to evaluate, selected depending on the acquisition function, is too close to one of the previous observations, it might lead to the computational issue reported in Chap. 3 that is the impossibility to compute the inverse of the matrix K and, consequently, make inference through the GP. The new acquisition function proposed in Yan et al. (2018) consists in the minimization of the condition number of K, leading to a method named Sequentially Bayesian K-optimal (SBKO) design, which tries to avoid sampling close to previous observations. Therefore, SBKO naturally forces the samples to spread sparsely in the search space, providing an alternative exploration mechanism. A possible integration of K-optimality into BO is to consider one of the possible acquisition functions, such as EI, and then select, among possible candidate points with similar values of acquisition the one with smaller condition number. However, this kind of approach leaves some cases undecided. More precisely, given two candidate points, x' and x'', it is not clear which select in the cases: (a) x' is better than x'' in terms of acquisition function but has a higher condition number or (b) x' is worse than x'' in terms of acquisition function but has a lower condition number. To solve these cases, Yan et al. (2018) proposes to consider the equation of parametrized EI (reported in Sect. 4.1.2):

$$\mathrm{EI}(x) = \begin{cases} \left(f\left(x^{+}\right) - \mu(x) - \xi\right)\Phi(Z) + \sigma(x)\phi(Z) \text{ if } \sigma(x) > 0 \\ 0 \text{ if } \sigma(x) = 0 \end{cases}$$

where ϕ and Φ are the probability distribution and the cumulative distribution functions, respectively, and

$$Z = \begin{cases} \frac{f(x^{+}) - \mu(x) - \xi}{\sigma(x)} \text{ if } \sigma(x) > 0 \\ 0 \text{ if } \sigma(x) = 0 \end{cases}$$

The approach consists in making the parameter ξ dependent on the condition number κ through:

$$\xi(\kappa) = \frac{\log \kappa}{\log \kappa + c \log \kappa_T}$$

where κ_T is a suitable threshold of the condition number and c is a constant that controls the shape of $\xi(\kappa)$. The approach addresses an important issue in BO but numerical evidence is still partial.

4.3 Optimizing the Acquisition Function

The optimization of acquisition functions has been receiving comparably less attention than the development of new ones and in general of the other components of the GP-BO machinery. Indeed, the traditional acquisition functions are available analytically and the computational cost of their optimization is usually much smaller than a single evaluation of the objective function: still the optimization of the acquisition function plays a crucial role as it suggests the next point and it is quite difficult anyway, in high-dimensional spaces, given their non-convexity and specific shape features.

Commonly adopted techniques are random multi-start with local searches or genetic algorithms. More precisely, taking into account the BO software and platforms presented in Chap. 6, genetic algorithm (GA) and evolutionary algorithm (EA) are the most widely adopted techniques to optimize the acquisition function. DIRECT is also another common option; BFGS and L-BFGS implementations can be also adopted in the case of a continuous probabilistic surrogate model, such as GP.

A recent paper Wilson et al. (2018) takes a new approach to the optimization of the acquisition function, looking especially at two different topics:

1. The gradient of acquisition function is estimated via Monte Carlo integration and then used in gradient-based optimization.
2. Greedy maximization of myopic acquisition function is shown, using submodularity, to converge with a guaranteed accuracy.

In the first topic, they use infinitesimal perturbation analysis (Glasserman 1988), also proposed in Kingma and Welling (2013) under the name of *reparametrization trick*. The second topic is concerned with the family of myopic maximization (MM) functions, defined as the expected max of a point-wise utility function: $MM(\mathbf{X}) = \mathbb{E}_{\mathbf{y}}\left[\alpha(\mathbf{y})\right]$.

The family of MM functions is submodular (SM) and greedly maximizing SM functions are guaranteed to produce near-optimal results Krause and Golovin (2014).

References

Astudillo, R., Frazier, P.: Bayesian optimization of composite functions. In: International Conference on Machine Learning, pp. 354–363 (2019, May)

Auer, P.: Using confidence bounds for exploitation-exploration trade-offs. J. Mach. Learn. Res. **3**(3), 397–422 (2002)

Basu, K., Ghosh, S.: Analysis of Thompson sampling for Gaussian process optimization in the bandit setting (2017). arXiv preprint arXiv:1705.06808

Berger, J.O.: Statistical Decision Theory and Bayesian Analysis. Springer Science & Business Media (2013)

Brochu, E., Cora, V.M., de Freitas, N.: A tutorial on bayesian optimization of expensive cost functions, with application to active user modeling and hierarchical reinforcement learning (2010). arXiv preprint arXiv:1012.2599

Frazier, P., Powell, W., Dayanik, S.: The knowledge-gradient policy for correlated normal beliefs. Informs J. Comput. **21**(4), 599–613 (2009)

Glasserman, P.: Performance continuity and differentiability in Monte Carlo optimization. In: 1988 Winter Simulation Conference Proceedings, pp. 518–524. IEEE (1988)

González, J., Osborne, M., Lawrence, N.D.: GLASSES: Relieving the Myopia of Bayesian Optimisation (2016)

Hennig, P., Schuler, C.J.: Entropy search for information-efficient global optimization. J. Mach. Learn. Res. **13**, 1809–1837 (2012)

Hernández-Lobato, J.M., Hoffman, M.W., Ghahramani, Z.: Predictive entropy search for efficient global optimization of black-box functions. Adv. Neural. Inf. Process. Syst. **25**, 144–149 (2014). https://doi.org/10.1016/j.molstruc.2009.06.011

Jones, D.R., Schonlau, M., Welch, W.J.: Efficient global optimization of expensive black-box functions. J. Glob. Optim. **13**(4), 455–492 (1998)

Kingma, D.P., Welling, M.: Auto-encoding variational bayes (2013). arXiv preprint arXiv:1312.6114

Krause, A., Golovin, D.:. Submodular function maximization. In: Tractability: practical Approaches to Hard Problems, pp. 71–104. Cambridge University Press (2014)

Kushner, H.J.: A new method of locating the maximum point of an arbitrary multi-peak curve in the presence of noise. J. Basic Eng. **86**, 97–106 (1964)

Lam, R., Willcox, K., Wolpert, D.H.: Bayesian optimization with a finite budget: an approximate dynamic programming approach. Adv. Neural. Inf. Process. Syst. **30**, 883–891 (2016). https://doi.org/10.1186/1471-2407-4-76

Marchant, R., Ramos, F., Sanner, S.: Sequential Bayesian optimisation for spatial-temporal monitoring. In: International Conference on Uncertainty in Artificial Intelligence, pp. 553–562 (2014)

Mockus, J., Tiesis, V., and Zilinskas, A.: The application of Bayesian methods for seeking the extremum. In: Dixon, L., Szego, G. (eds.) Towards Global Optimisation, 2, pp. 117–130. Elsevier (1978)

Nguyen, V., Osborne, M.A.: Knowing the what but not the where in Bayesian optimization (2019). arXiv preprint arXiv:1905.02685

Noè, U., Husmeier, D.: On a new improvement-based acquisition function for Bayesian optimization (2018). arXiv preprint arXiv:1808.06918

Osborne, M.A., Garnett, R., Roberts, S.J.: Gaussian processes for global optimization. In: 3rd International Conference on Learning and Intelligent Optimization (LION3), vol. 2009 (2009)

Poloczek, M., Wang, J., Frazier, P.: Multi-information source optimization.In: Guyon, I., Luxburg, U.V., Bengio, S., Wallach, H., Fergus, R., Vishwanathan, S., Garnett, R. (eds.) Advances in Neural Information Processing Systems, 30, pp. 4291–4301. Curran Associates, Red Hook, NY (2017)

Russo, D., Van Roy, B., Kazerouni, A., Osband, I., Wen, Z.: A tutorial on Thompson sam-pling. Found. Trends Mach. Learn. **11**, 1–96 (2018). https://doi.org/10.1561/2200000070

Srinivas, N., Krause, A., Kakade, S.M., Seeger, M.W.: Information-theoretic regret bounds for Gaussian process optimization in the bandit setting. In: IEEE Transactions on Information Theory, pp. 3250–3265 (2012)

Toscano-Palmerin, S., Frazier, P.I.: Bayesian optimization with expensive integrands (2018). arXiv preprint arXiv:1803.08661

Villemonteix, J., Vazquez, E., Walter, E.: An informational approach to the global optimization of expensive-to-evaluate functions. J. Global Optim. **44**(4), 509 (2009)

Volpp, M., Fröhlich, L., Doerr, A., Hutter, F., Daniel, C.: Meta-learning acquisition functions for Bayesian optimization (2019). arXiv preprint arXiv:1904.02642

Wang, Z., Jegelka, S.: Max-value entropy search for efficient Bayesian optimization. In: Proceedings of the 34th International Conference on Machine Learning. Sydney, Australia (2017)

Wilson, J., Hutter, F., Deisenroth, M.: Maximizing acquisition functions for Bayesian optimization. In: Advances in Neural Information Processing Systems, pp. 9906–9917 (2018)

Wu, J., Frazier, P.: The parallel knowledge gradient method for batch bayesian optimization. In: Advances in Neural Information Processing Systems, pp. 3126–3134 (2016)

Wu, J., Frazier, P.I.: Discretization-free knowledge gradient methods for bayesian optimization (2017). arXiv preprint arXiv:1707.06541

Wu, J., Poloczek, M., Wilson, A.G., Frazier, P.: Bayesian optimization with gradients. In: Advances in Neural Information Processing Systems, pp. 5273–5284, (4.2.5) (2017)

Yan, L., Duan, X., Liu, B., Xu, J.: Bayesian optimization based on K-optimality. Entropy **20**(8), 594 (2018)

Chapter 5
Exotic Bayesian Optimization

5.1 Constrained Global Optimization

GO and BO have been considered first in "essentially unconstrained" conditions, where the solution was searched for within a bounded-box search space. Recently, due to methodological and application reasons, there has been an increasing interest in constrained global optimization (CGO).

$$x^* = \arg\min_{x \in X} f(x)$$
$$\text{Subject to}$$
$$c_i(x) \leq 0\, i = 1, \ldots, n_c$$

A paper of general interest about a taxonomy of constraints in simulation-based optimization is given in Le Digabel (2015). Some general remarks have to be taken into account:

1. Even if the constraints are analytically defined, they presence without the convexity assumption raises a whole new set of challenging topics, in particular, the interaction between the feasible region and the surrogate probabilistic model.
2. Since BO assumes the objective function as black box, a natural extension is to consider the constraints as black box as well (unknown constraints).
3. A relevant case is when the objective function is undefined, and cannot be therefore computed, outside the feasible region. In this case, we speak about partially defined objective functions (Rudenko 1994; Sergeyev et al. 2007).

Point three is particularly relevant when the evaluation of the objective requires, as in black box conditions, the execution of a simulation model which can return a valid output or a failure message. An example of failure can be a computational fluid dynamics solver that does not converge due to instability of the numerical scheme (Sacher et al. 2018) or a hydraulic simulator if the input generates pressures or flows physically impossible (Tsai et al. 2018). Also frequent is the case of a

F. Archetti and A. Candelieri, *Bayesian Optimization and Data Science*, SpringerBriefs in Optimization, https://doi.org/10.1007/978-3-030-24494-1_5

complex machine learning model where the training of the model, and therefore the evaluation of the loss function, relies on gradient (stochastic) descent which may fail to converge. Such an event prevents the exploration of a neighbourhood of a "not computable" point and halts the sequential optimization procedure. A naïve solution, in presence of a not computable point, is to associate to it a fixed high (low) penalty value for the objective function to be minimized (maximized): still, determining suitable value is not a trivial task and might imply, anyway, a loss of accuracy in the Gaussian surrogate model and of sample efficiency.

There have been several attempts at leveraging BO framework into dealing with constrained optimization: the main problem is to propose an acquisition function for CBO.

The use of GP-and EI-based heuristics has been first proposed in Jones et al. (1998) allowing for $c_i(x)$ to be black box and assuming their mutual independence and with the objective function. A GP is given as a prior to each constraint. If f_c^+ is the best feasible observation of f, the EI acquisition function is:

$$\mathrm{EI}(x|f_c^+) = \begin{cases} (\mu(x) - f_c^+)\Phi(Z) + \sigma(x)\phi(Z) \text{ if } \sigma(x) > 0 \\ 0 \text{ if } \sigma(x) = 0 \end{cases}$$

where ϕ and Φ are the probability distribution and the cumulative distribution functions, respectively, and

$$Z = \begin{cases} \frac{\mu(x)-f_c^+}{\sigma(x)} \text{ if } \sigma(x) > 0 \\ 0 \text{ if } \sigma(x) = 0 \end{cases}$$

In presence of constraints the formula becomes:

$$\mathrm{EIC}(x|f_c^+) = \mathrm{EI}(x|f_c^+)\prod_{i=1}^{n_c}\mathbb{P}(c_i(x) \leq 0)$$

where the improvement of a candidate solution x over f is zero if x is not feasible. When the noise in the constraints is taken into account, we may not know which observations are feasible: for instance, the best GP mean value satisfying each constraint $c_i(x)$ with a probability at least $1 - \delta_i$ can be used (Letham et al. 2019). Gramacy (2016) proposes a different approach for handling constraints in which they are brought into the objective function via a Lagrangian. EI is no longer tractable analytically but can be evaluated numerically via Monte Carlo integration or quadrature (Picheny 2016). Relevant prior results on BO with unknown constraints proposed new acquisition functions, such as integrated expected conditioned improvement (IECI) (Gramacy and Lee 2011). Another approach is presented in Feliot et al. (2017) where an adaptive Random Search is used to approximate feasible region while optimizing the objective function. A penalty approach has been considered first in Gardner et al. (2014), where a penalty is assigned directly to the acquisition function in case of infeasibility, with the aim to move away from infeasible regions. A similar approach has been extended in Candelieri et al. (2018) also in the case in which the function is partially defined: infeasibility is treated by assigning a fixed penalty as value of the objective function (we refer as "BO with penalty").

A new general approach is offered by information-based methods, which have been extended to the constrained case (e.g. predictive entropy search with constraints, PESC) in Hernandez-Lobato et al. (2015): the code for PESC is included in Spearmint and available at https://github.com/HIPS/Spearmint/tree/PESC.

The above approaches assume that the number of constraints is known a priori and they are statistically independent.

The assumption of independence permits to compute the probability of feasibility simply as the product of individual probabilities with respect to every constraint. The result is multiplied by the acquisition function whose optimization would prefer points satisfying the constraints with high probability.

The issue of partially defined function or equivalently "crash constraints" or "non-computable domains" transforms the GP-based feasibility evaluation from a regression problem into a classification one. This classification approach has been analyzed in several papers and discussed in the following.

In Basudhar et al. (2012) a probabilistic SVM (PSVM) is used to calculate the so-called probability of feasibility and the optimization scheme alternates between a global search for the optimal solution, depending on both this probability and the estimated value of the objective function—modelled through a GP—and a local refinement of the PSVM through an adaptive local sampling scheme. The most common PSVM model is based on the sigmoid function (Vapnik 1998). For a given sample x, the probability of belonging to the $+1$ class (i.e. "feasible") is:

$$P(+1|x) = \frac{1}{1 + e^{As(x)+B}}$$

The parameters A ($A < 0$) and B of the sigmoid function are found by maximum likelihood. Two alternative formulations of the acquisition function are proposed:

$$\max_x \mathrm{EI}(x) P(+1|x)$$

and

$$\begin{aligned} &\max_x \mathrm{EI}(x) \\ &\text{Subject to} \\ &P(+1|x) \geq 0.5 \end{aligned}$$

Other approaches, also based on GP, and focused on cases where the objective function cannot be computed (not computable domains or crash constraints) have been proposed in Sacher et al. (2018); Bachoc et al. (2019).

The previous approaches use independent GPs to model the objective function and the constraints, requiring two strong assumptions: a priori knowledge about the number of constraints and the independence among objective function and all the constraints. Extending EI to account for correlations between constraints and the objective function is still, according to Letham et al. (2019), an open challenge.

To overcome this limitation, a new approach, namely SVM-CBO (Support Vector Machine based Constrained BO), has been proposed in Candelieri and Archetti (2019): the main contribution of the paper is the development of a method which does not require any of the two previous assumptions. The approach uses support vector machine (SVM) to sequentially estimate and model the unknown feasible region (Ω) within the search space (i.e. feasibility determination), without any assumption on the number of constraints as well as their independence.

SVM-CBO is organized in two phases: the first is aimed to provide a first estimate of Ω (feasibility determination) and the second is BO performed on such an estimate, only. The motivation is that we are interested in obtaining a good approximation of the overall feasible region and not only close to the optimal solution (i.e. feasibility determination is a goal per se, in our approach). Another relevant difference with Basudhar et al. (2012) is that SVM-CBO uses more efficiently the available "budget" (i.e. maximum number of function evaluations): at every iteration, of both phase 1 and 2, we perform just one function evaluation, while in the boundary refinement of Basudar et al. (2012) a given number $\boldsymbol{n_p}$ of function evaluations is performed in the neighbourhood of the next point to evaluate, with the aim to locally refine the boundary estimate. SVM-CBO is detailed in the following section.

5.2 Support Vector Machine—Constrained Bayesian Optimization

We start with the definition of the problem, that is:

$$\min_{x\in\Omega\subset X\subset\mathbb{R}^d} f(x)$$

where $f(x)$ has the following properties: it is black box, multi-extremal, expensive and partially defined. The last feature means that $f(x)$ is undefined outside that feasible region Ω, which is a subset of overall bounded-box search space $X \subset \mathbb{R}^d$. Moreover, we consider the case that constraints defining the feasible region are also black box.

We introduce some notation that will be used in the following:

- $D_n^\Omega = \{(x_i, y_i)\}_{i=1,\ldots,n}$ is the **feasibility determination** dataset;
- $D_l^f = \{(x_i, f(x_i))\}_{i=1,\ldots,l}$ is the **function evaluations** dataset, with $l \le n$ (because $f(x)$ is partially defined on X) and where l is the number of points where it was possible to compute f out of the n queried so far;

and x_i is the i-th queried point and $y_i = \{+1, -1\}$ defines if x_i is feasible or infeasible, respectively.

Phase 1—Feasibility determination

The first phase of the approach aims to find an estimate $\tilde{\Omega}$ of the actual feasible region Ω in M function evaluations ($\tilde{\Omega}_M = \tilde{\Omega}$). The sequence of function evaluations is determined according to an SMBO process where the surrogate model—of the feasible region, in this phase—provides the currently estimated feasible region $\tilde{\Omega}_n$. As surrogate model we use the (non-linear) separation hyperplane of an SVM classifier, trained on the set D_n^{Ω}. The SVM classifier uses an RBF kernel to model feasible regions with non-linear boundaries.

Let denote with $h_n(x)$ the argument of the SVM-based classification function:

$$h_n(x) = \sum_{i=1}^{n_{SV}} \alpha_i y_i k(\bar{x}_i, x) + b$$

where α_i and y_i are the Lagrangian coefficient and the "feasibility label" of the i-th support vector, $\bar{x}_i$, respectively, $k(.,.)$ is the kernel function (i.e. an RBF kernel, in this study), b is the offset and n_{SV} is the number of support vectors.

The boundaries of the estimated feasible region $\tilde{\Omega}_n$ are given by $h_n(x) = 0$ (i.e. non-linear separation hyperplane). The SVM-based classification function provides the estimated feasibility for any $x \in$ X:

$$\tilde{y} = \text{sign}(h_n(x)) = \begin{cases} +1 \text{ if} x \in \tilde{\Omega}_n \\ -1 \text{ if} x \notin \tilde{\Omega}_n \end{cases}$$

With respect to the aim of the first phase, we propose a *"feasibility acquisition function"* aimed at identifying the next promising point according to two different goals:

- Improving the estimate of feasible region
- Discovering possible disconnected feasible regions.

To deal with the first goal, we use the distance of x from the boundaries of the currently estimated feasible region $\tilde{\Omega}_n$, using the following formula from the SVM classification theory:

$$d_n(h_n(x), x) = |h_n(x)| = \left| \sum_{i=1}^{n_{SV}} \alpha_i y_i k(\bar{x}_i, x) + b \right|$$

To deal with the second goal, we introduce the concept of "coverage of the search space", defined by:

$$c_n(x) = \sum_{i=1}^{n} e^{-\frac{\| \bar{x}_i - x^2 \|}{2\sigma_c^2}}$$

So, $c_n(x)$ is a sum of n RBF functions centred on the points evaluated so far, with σ_c a parameter to set the width of the corresponding bell-shaped curve.

Finally, the *feasibility acquisition function* is given by the sum of $d_n(h_n(x), x)$ and $c_n(x)$, and the next promising point is identified by solving the following optimization problem:

$$x_{n+1} = \underset{x \in X}{\operatorname{argmin}}\{d_n(h_n(x), x) + c_n(x)\}$$

Thus, we want to select the point associated with the minimal distance from the boundaries of the current estimated feasible region and the minimal coverage (i.e. max uncertainty). This allows us to balance between improving the estimate of the feasible region and discovering possible disconnected feasible regions (in less explored areas of the search space). It is important to highlight that, in phase 1, the optimization is performed on the overall bounded-box search space X.

After the function evaluation of the new point x_{n+1}, the following information is available:

$$y_{n+1} = \begin{cases} +1 \text{ if } x_{n+1} \in \Omega;\ f(x_{n+1}) \text{ is defined} \\ -1 \text{ if } x_{n+1} \notin \Omega : f(x_{n+1}) \text{is not defined} \end{cases}$$

and the following updates are performed:

- Feasibility determination dataset and estimated feasible region $\tilde{\Omega}_{n+1}$

$$\begin{gathered} D_{n+1}^{\Omega} = D_n^{\Omega} \cup \{(x_{n+1}, y_{n+1})\} \\ h_{n+1}(x) | D_{n+1}^{\Omega} \\ n \leftarrow n + 1 \end{gathered}$$

- Only if $x \in \Omega$, function evaluations dataset

$$\begin{gathered} D_{l+1}^{f} = D_l^{f} \cup \{(x_{l+1}, f(x_{l+1}))\} \\ l \leftarrow l + 1 \end{gathered}$$

The SMBO process for phase 1 is repeated until $n = M$.

Phase 2—Bayesian Optimization in the estimated feasible region

In this phase, a traditional BO process is performed but with the following relevant differences:

- the search space is not box-bounded but the estimated feasible region $\tilde{\Omega}_n$ identified in phase 1

- the surrogate model—a GP—is fitted only using the feasible solutions observed so far, D_l^f
- the acquisition function for phase 2—lower confidence bound (LCB), in this study—is defined on $\tilde{\Omega}_n$, only

Thus, the next point to evaluate is given by:

$$x_{n+1} = \underset{x \in \tilde{\Omega}_n}{\text{argmin}}\{\text{LCB}_n(x) = \mu_n(x) - \beta_n \sigma_n(x)\}$$

where $\mu_n(x)$ and $\sigma_n(x)$ are the mean and the standard deviation of the current GP-based surrogate model and β_n is the inflate parameter to deal with the trade-off between exploration and exploitation for this phase. It is important to highlight that, contrary to phase 1, the acquisition function is here minimized on $\tilde{\Omega}_n$, only, instead of the entire bounded-box search domain X.

The point x_{n+1} is just expected to be feasible, according to $\tilde{\Omega}_n$ but the information on its actual feasibility is known only after having checked whether $f(x_{n+1})$ can or cannot be computed (i.e. it is defined or not in x_{n+1}). Subsequently, the feasibility determination dataset is updated as follows:

$$D_{n+1}^{\Omega} = D_n^{\Omega} \cup \{(x_{n+1}, y_{n+1})\}$$

and according to the two alternative cases:

- x_{n+1} is actually feasible: $x_{n+1} \in \Omega$, $y_{n+1} = +1$;

the function evaluations dataset is updated as follows: $D_{l+1}^f = D_l^f \cup \{(x_{l+1}, f(x_{l+1}))\}$, with $l \leq n$ is the number of the feasible solutions with respect to all the points observed. The current estimated feasible region $\tilde{\Omega}_n$ can be considered accurate and retraining of the SVM classifier can be avoided: $\tilde{\Omega}_{n+1} = \tilde{\Omega}_n$

- x_{n+1} is actually infeasible: $x_{n+1} \notin \Omega$, $y_{n+1} = -1$;

the estimated feasible region must be updated to reduce the risk for further infeasible evaluations

$$h_{n+1}(x)|D_{l+1}^f \Rightarrow \tilde{\Omega}_{n+1}$$

The phase 2 continues until the overall available budget $n = N$ is reached.

In Candelieri (2019) the SVM-CBO approach has been validated on five 2D test functions for CGO and compared to "BO with penalty". An overall budget of 100 function evaluations, divided as follows:

- 10 for initialization through Latin hypercube sampling (LHS);
- 60 for feasibility estimation (phase 1)
- and 30 for SMBO constrained to estimated feasible region (phase 2).

In the case of BO with penalty, the same budget has been divided as follows:

- 10 evaluations for initialization (LHS)
- and 90 for BO (that is the sum of budget used for phase 1 and phase 2 in the proposed approach).

For each independent run, the initial set of solutions identified through LHS is the same for SVM-CBO and BO with penalty, in order to avoid differences in the values of the gap metric due to different initialization.

The so-called *Gap metric* has been used to measure the improvement obtained along the SMBO process with respect to global optimum $f(x^*)$ and the initial best solution $f(x_0)$ obtained from the initialization step:

$$G_n = \frac{|f(x_0) - f(x^+)|}{|f(x_0) - f(x^*)|}$$

where $f(x^+)$ is the "best seen" up to iteration n. Gap metrics varies in the range [0, 1]. For statistical significance, the gap metrics have been computed on 30 different runs, performed for every test function and for both SVM-CBO and BO with penalty.

While gap metric allows for comparing SVM-CBO to BO with penalty, it was important to introduce another performance measure to quantify how good is the approximation of the feasible region, along the SMBO process. We have defined a simple *overlap metric* as follows:

$$O_n\left(\Omega, \tilde{\Omega}_n\right) = \begin{cases} \frac{\text{Vol}(\tilde{\Omega}_n \bigcap \Omega)}{\text{Vol}(\Omega)} \text{if Vol}\left(\tilde{\Omega}_n\right) < \text{Vol}(\Omega) \bigvee \left(\text{Vol}\left(\tilde{\Omega}_n \bigcap \Omega\right)\right. \\ \left.= \text{Vol}(\Omega) \bigwedge \text{Vol}\left(\tilde{\Omega}_n\right) = \text{Vol}(\Omega)\right) \\ \frac{\text{Vol}(\tilde{\Omega}_n)}{\text{Vol}(\Omega)} \text{ otherwise} \end{cases}$$

where Ω and $\tilde{\Omega}_n$ are the actual and the estimated feasible regions, respectively, and Vol() is the volume of a region. More precisely, the volume of a region is computed as an approximation by simply generating a grid of points within the search space and then counting the number of points falling into that region.

According to its definition, the overlap metric can vary in the range $[0, \infty)$, where a good approximation of the feasible region is associated with a value equal to 1 while no overlap is associated with a value equal to 0. Values higher than 1 represent situations where $\Omega \subset \tilde{\Omega}_n$.

The following set of figures show the gap metric computed for every test function with respect to the number of function evaluations, excluding the initialization through LHS. The value of gap metric at iteration 0 is the best seen at the end of the initialization step (i.e. $f(x_0)$ in the gap metric formula). Each graph compares the gap metric provided by SVM-CBO and BO with penalty, respectively. Both the average and the standard deviation of the gap metric, computed on 30 independent runs for each approach, are depicted. The higher effectiveness of the proposed approach is clear, even considering the variance in the performances. The end of phase 1 is represented by a vertical dotted line in the charts; a significant improvement of the

SVM-CBO's gap metric is observed after this phase, in every test case. It is important to remind that phase 1 of the SVM-CBO is aimed at approximating the unknown feasible region, while the optimization process only starts with phase 2. Thus, the relevant shift in the gap metric is motivated by the explicit model of the feasible region learned in phase 1 (Fig. 5.1).

Finally, the following set of figures show how the *overlap metric* changes sequentially over the SMBO process. Since 10 initial function evaluations are all used to provide a first estimate of the feasible region, the x-axis is limited to 90, that is the portion of the budget allocated for phase 1 and phase 2: during these phases—particularly during phase 1—the estimate of the feasible region is updated and the overlap metric changes consequently. The dotted line represents the mean overlap

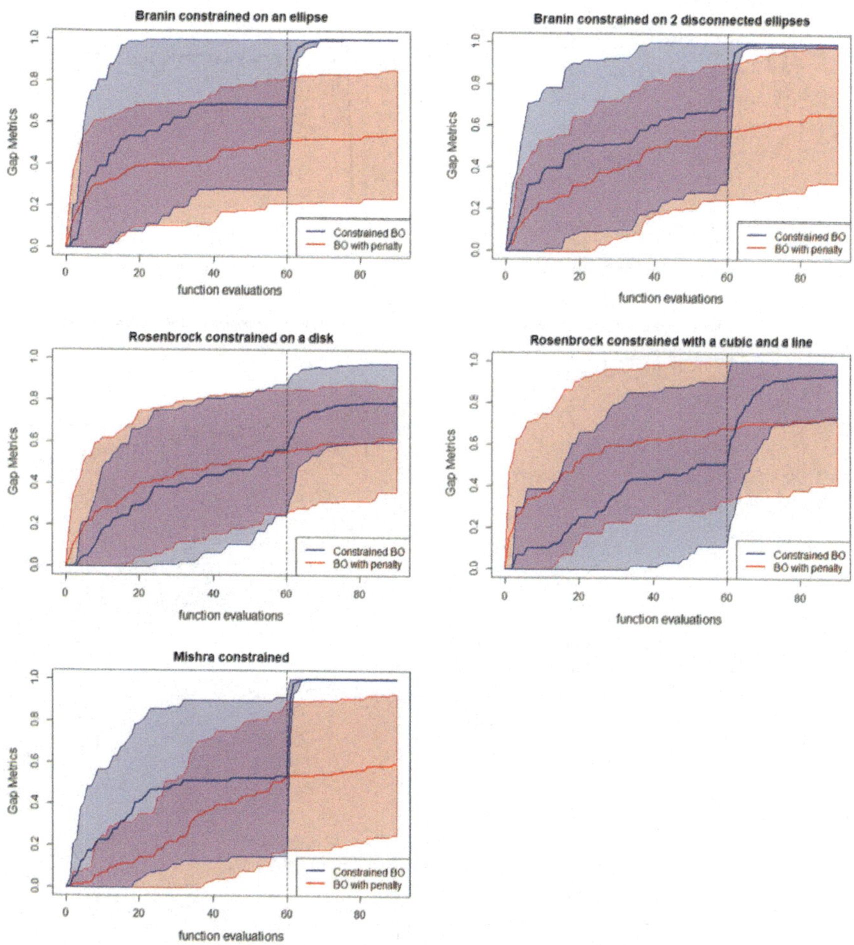

Fig. 5.1 Comparison of gap metrics for the proposed SVM-CBO approach versus BO with penalty, for each one of the five test functions, respectively

metric and the shaded area represents the standard deviation. The figures show that the proposed SVM-CBO approach achieves a reasonable overlap (higher than 77%) between estimated and actual feasible region, already after 30–35 function evaluations. The most difficult case was the one with a feasible region consisting of two disconnected ellipses, but, before the end of budget, overlap metric achieved anyway the value of 80% (Fig. 5.2).

SVM-CBO offers a good estimation of the feasible search space, even for complex feasible search spaces consisting of disconnected regions. Seldom, the BO process, in phase 2, suggested infeasible points but, thanks to the update of the estimated feasible region, it was able to quickly fall close to the optimum, without wasting

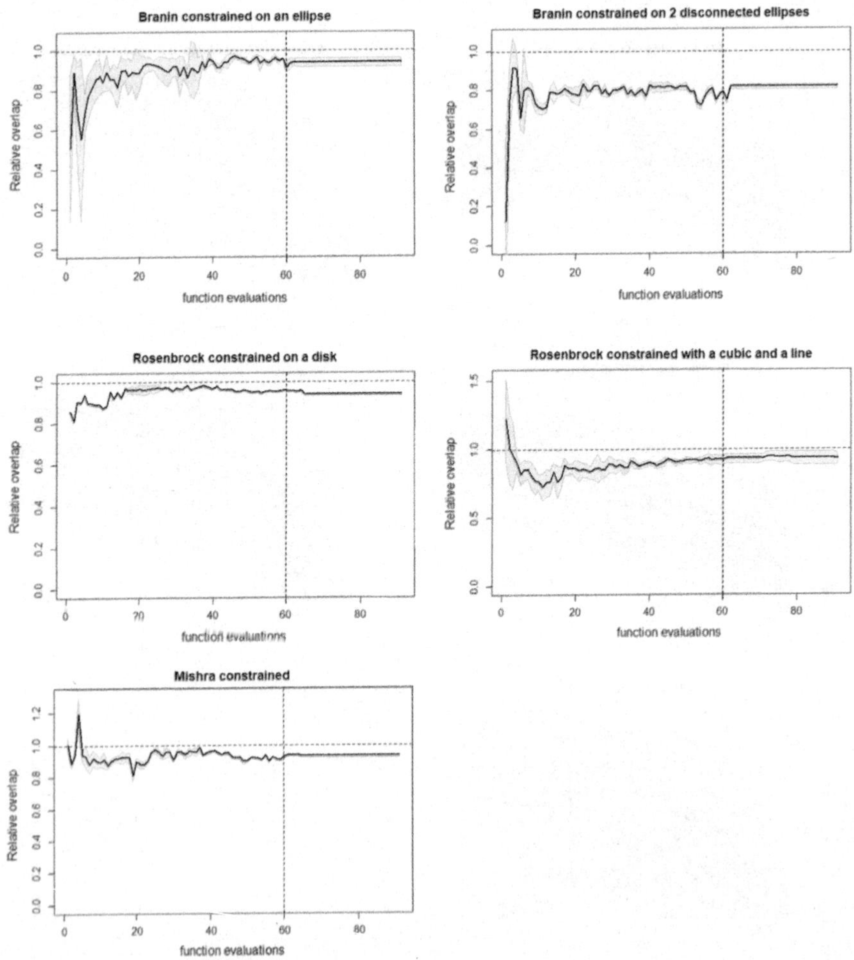

Fig. 5.2 Overlap metric measuring the quality of feasibility estimation provided by the proposed approach, for each one of the five test functions, respectively

further function evaluations outside the actual feasible region. Other important considerations relative to significant reduction in computational costs are reported in Candelieri (2019).

In the following, some explicative pictures relative to how SVM-CBO works are reported, considering a 2D test function (i.e. Branin) where the unknown feasible region consists of the inner areas within two disconnected ellipses.

Begin of phase 1: first estimation of the feasible region according to 10 functions evaluations (i.e. initialization through LHS). The following symbols have been used:

- Solid black line: boundaries of the feasible region (feasible points are inside the ellipses)
- Dotted blue line: boundaries of the estimated feasible region
- Green points: optima of (original) Branin's test function; only one of them is feasible!
- Blue/black points: evaluations resulting feasible/infeasible
- Points within diamonds—black or blue: support vectors of the two classes
- Red-circled points—black or blue: classification errors with respect to the two classes
- Red asterisk: next point to evaluate.

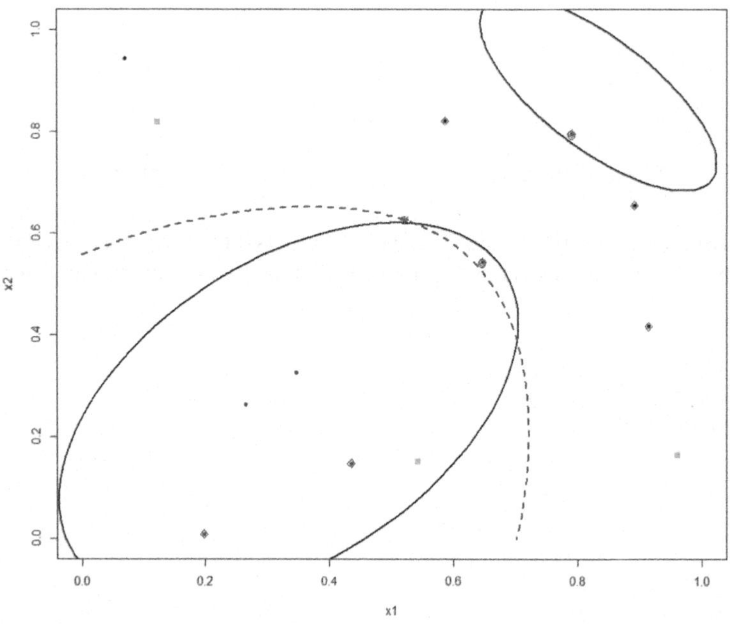

Following, modifications of the estimated feasible region along with some iterations (i.e. 1, 5, 9 and 10) of phase 1:

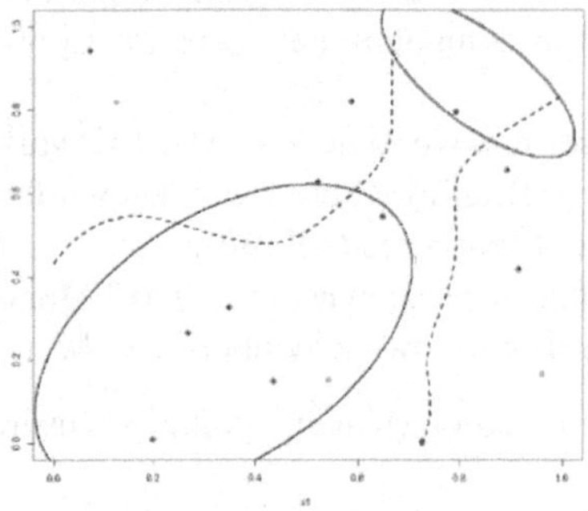

Phase 1 - Iteration #1

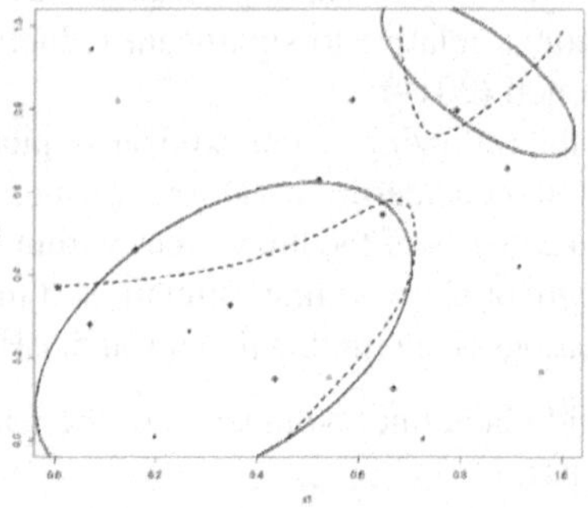

Phase 1 - Iteration #5

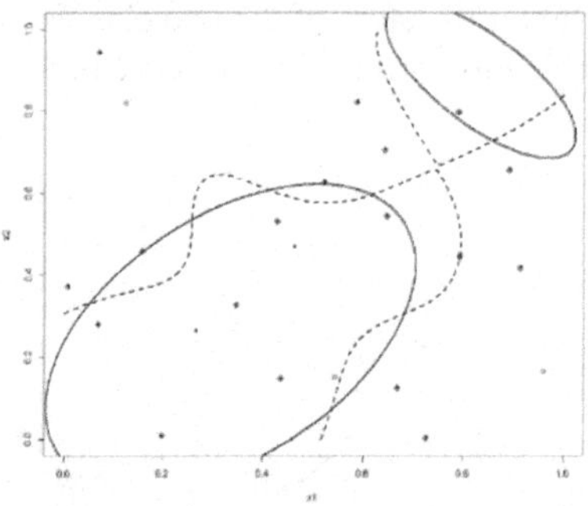

Phase 1 - Iteration #9

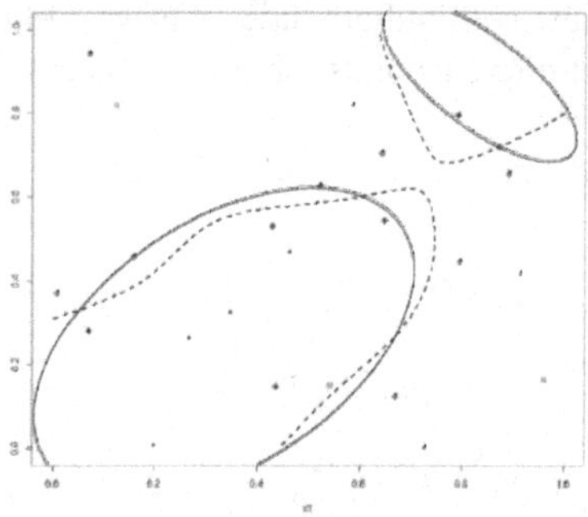

Phase 1 - Iteration #10

End of Phase 1 and beginning of Phase 2: level sets for the original box-constrained objective function are also depicted. In the following figure, the estimated and actual feasible regions are reported.

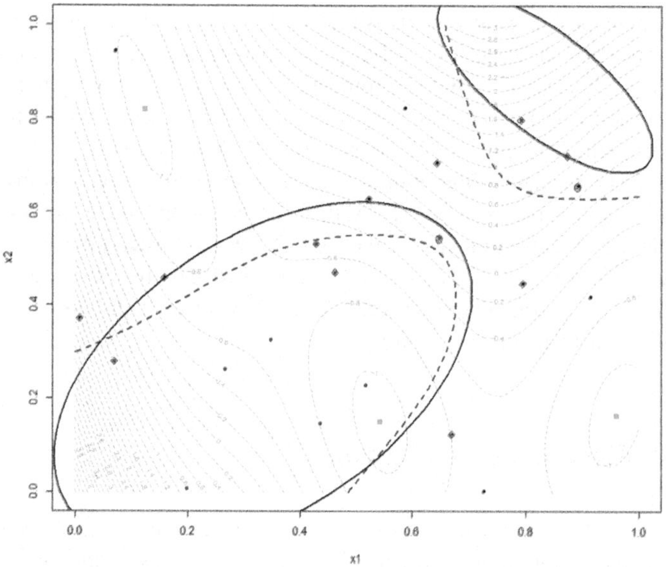

End of Phase 2 (10 function evaluations more): blue triangles are function evaluations performed in phase 2. All the triangles are in the estimation of the feasible region obtained at the end of phase 1. This means that the estimated feasible region has not been updated during phase 2. The feasible global optimum has been identified in phase 2.

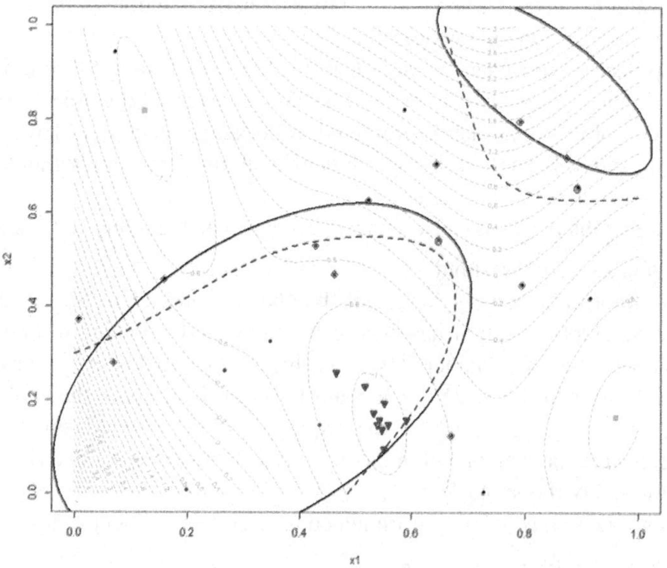

SVM-CBO has some limitations which can be relevant in some application context. First, as any other constrained global optimization approach, it is not well-suited for dealing with problems where infeasible evaluations are "destructive". This means that cannot be applied—as is—for solving optimal (online) control of complex industrial systems. This specific kind of problems, known as safe exploration–optimization, will be addressed in the following section. Therefore, simulation–optimization is the setting which might largely benefit from SVM-CBO, such as demonstrated in the case of pump scheduling optimization in water distribution network by using the hydraulic simulation software EPANET 2.0. Secondly, since the SVM classifier models the overall boundary of the feasible region, instead of each individual constraint, a sensitivity analysis cannot be performed. So, the advantage—not only computational—of using one SVM could be partially offset, in the case, we are facing a problem requiring sensitivity analysis.

5.3 Safe Bayesian Optimization

The important difference with respect to CGO is that SafeOpt does not allow—at least with a given probability—any function evaluations outside the feasible region. Moreover, in SafeOpt the objective function is not necessarily partially defined: indeed, a function evaluation is unsafe if the corresponding value of the objective function violates the safety threshold.

It is important to highlight that: contrary to the rest of this book, where the global optimization problem is defined as a minimization problem, for safe optimization we consider a maximization problem, to be coherent with the literature on this topic. Thus, we want to maximize some black-box expensive function without performing function evaluations under a given value.

An approach is presented in Sui et al. (2015), which tries to optimize taking into account the—supposedly known—Lipschitz constant of the objective function. An extension of this approach is presented in Berkenkamp et al. (2016), where the SafeOpt algorithm is applied on high-dimensional problems and without using the information on Lipschitz constant.

The basic problem was stated as: $\max_{x \in X} f(x)$, subject to the safety constraint $f(x) \geq h$, where h is a safety threshold.

A new algorithm, namely StageOpt, has been proposed in Sui et al. (2018), where the SafeOpt algorithm has been generalized and extended to be more efficient and applicable to a broader class of problems. The goal is to optimize a black-box objective function $f : D \to \mathbb{R}$ from noisy evaluations at the sample points $x_1, x_2, \ldots, x_n \in X$. Any point in the sequence, when sampled, must be "safe", which means that for each one of m unknown safety functions $g_i(x) : X \to \mathbb{R}$ it lies above some safety threshold $h_i \in \mathbb{R}$.

This optimization problem is formulated as $\max_{x \in X} f(x)$ subject to the safety constraints $g_i(x) \geq h_i$ for $i = 1, \ldots, m$.

This is a more general formalization of the original SafeOpt (Sui et al. 2015), which used only a simple threshold on the value of the objective function, leading to a single constraint $g(x) = f(x) \geq h$.

As in constrained optimization, GPs are used to model both the objective function and the constraints. An important assumption in StageOpt is that each safety function g_i is L_i-Lipschitz continuous, with respect to some metric on X. This assumption is usually satisfied by the most commonly used kernels (Srinivas et al. 2010; Sui et al. 2015). Additionally, at least one initial "seed" set of safe points must be given, denoted as $S_0 \subset X$.

Since safe optimization works by safely expanding S_0, it is not guaranteed to identify the global optimizer x^*, specifically in the case that the region around x^* is disconnected from S_0. The key component of SafeOpt is the *one-step reachability operator*:

$$R_\varepsilon(S) := S \cup \bigcap_i \{x \in X | \exists x' \in S, g_i(x') - \varepsilon - L_i d(x', x) \geq h_i\}$$

where S is the current set of safe points (initially S), d is a distance measure in $\mathbb{R}^d$ and ε is the absolute error in considering x as safe. This operator allows to identify the set of all the points estimated as safe depending on the previous evaluations of f on S. Then, given the maximum number of function evaluations, N, we can define the subset of X reachable after N iterations from the initial safe seed set S_0 as the following: $R_\varepsilon^N(S_0) := R_\varepsilon(R_\varepsilon \ldots (R_\varepsilon(S)) \ldots)$, N times. Thus, the optimization problem becomes:

$$x^* = \text{argmax}_{x \in R_\varepsilon^T(S_0)} f(x).$$

StageOpt separates the safe optimization problem into two stages: an exploration phase in which the safe region is iteratively expanded, followed by an optimization phase in which Bayesian optimization is applied within the safe region. Similarly to SVM-CBO, presented for constrained BO, the overall number of function evaluations, N is divided into N_1 for phase 1 and N_2 for phase 2, with $N = N_1 + N_2$. To map the uncertainty of the GP model at the generic iteration n, StageOpt used the confidence intervals:

$$Q_n^i(x) := [\mu_n^i(x) \pm \beta_n \sigma_n^i(x)]$$

where β_n defines the level of confidence. It is easy to note that they correspond to the upper and lower confidence bound of the GP. In the formula, superscripts index the corresponding safety functions, while subscripts index iterations, as usual. Then, to guarantee both safety and progress in safe region expansion, StageOpt uses the following confidence intervals $C_{n+1}^i(x) := C_n^i(x) \bigcap Q_n^i(x)$, with $C_0^i(x) := [h_i, \infty]$ so that C_{n+1}^i are sequentially contained in C_n^i for all $n = 0, \ldots, N$. Upper and lower bounds of C_n^i can be computed and denoted as u_n^i and l_n^i, respectively.

The first stage of StageOpt is safe region expansion. An increasing sequence of safe subsets $S_n \subseteq X$ is computed based on the confidence intervals of the GP posterior:

$$S_{n+1} = \bigcap_i \bigcup_{x \in S_n} \{x' \in X | l_{n+1}^i(x) - L_i d(x, x') \geq h_i\}$$

At each iteration, StageOpt computes a set of expanders points G_n while definition is based on the function:

$$e_n(x) := \left| \bigcap_i \{x' \in X \backslash S_n | u_n^i(x) - L_i d(x, x') \geq h_i\} \right|$$

which (optimistically) quantifies the potential enlargement of the current safe set after sample a new decision x. Then, G_n is defined as follows:

$$G_n = \{x \in S_n : e_n(x) > 0\}$$

Finally, at each iteration StageOpt selects the expander with the highest predictive uncertainty, given by $x_{n+1} = \underset{x \in G_n}{\operatorname{argmax}} u_n^i - l_n^i$.

The second stage of StageOpt is BO applied within the safe region identified at the end of the first stage. GP-UCB is used as acquisition function.

In Sui et al. (2018) is illustrated a graphical comparison of the behaviour of SafeOpt (Berkenkamp et al. 2016; Sui et al. 2015) and StageOpt starting from the same safe seed point. The figure is reported as follows (Fig. 5.3).

Initially, both algorithms select the same points, as they use the same definition of safe expansion. However, StageOpt selects noticeably better optimization points than SafeOpt due to the UCB criterion.

In real applications, it may be difficult to compute an accurate estimate of the Lipschitz constants, which may have an adverse effect on the definition of the safe region and its expansion dynamics. One might use one of the solutions for updating L, outlined in Chap. 2, like DIRECT or MultL (Sergeyev and Kvasov 2017).

Within the safe optimization framework, one can use a modified version of SafeOpt that defines safe points using only the GPs (Berkenkamp et al. 2016). This modification can be directly applied to StageOpt as well. This alternative algorithm is structured as follows:

1. a point $\bar{x}$ is considered safe if the lower confidence bound of each safety GPs lies above the respective threshold, that is $l_i(\bar{x}) > h_i$
2. a safe point x' is considered as an expander depending on an optimistic measure of the expansion from x'. More practically, for each constraint a "temporary" GP is trained on the observations performed so far and an unsafe point x which represents an "artificial" evaluation: the associated value of the constraint function

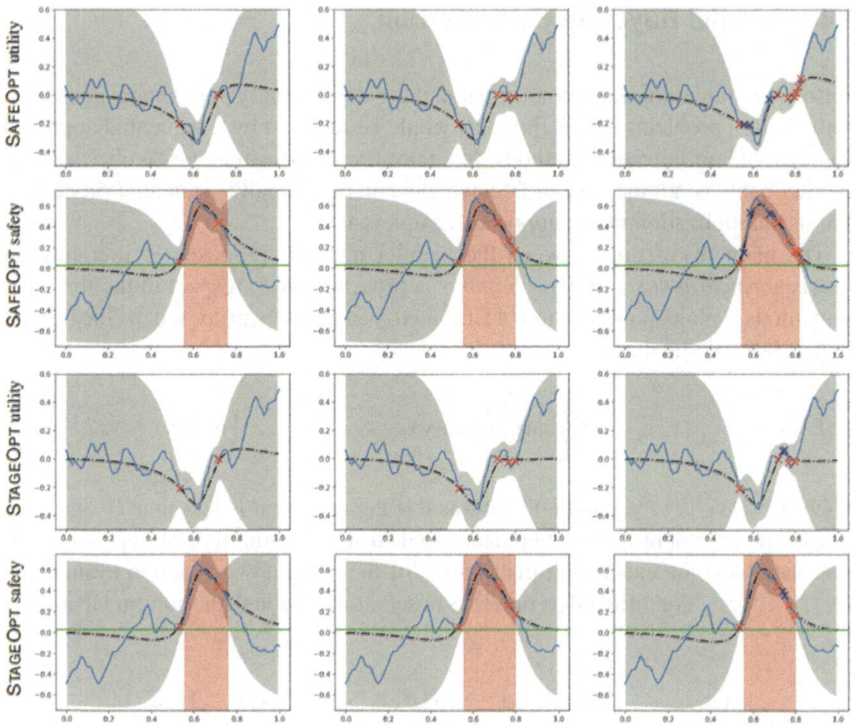

(a) 2 iterations (b) 4 iterations (c) 10 iterations

Fig. 5.3 Evolution of GPs in SafeOpt and StageOpt for a fixed safe seed; dashed lines correspond to the mean and shaded areas to ±2 standard deviations. The first and third rows depict the utility function, and the second and fourth rows depict a single safety function. The utility and safety functions were randomly sampled from a zero-mean GP with a Matern kernel and are represented with solid blue lines. The safety threshold is shown as the green line, and safe regions are shown in red. The red markers correspond to safe expansions and blue markers to maximizations and optimizations. We see that StageOpt identifies actions with higher utility than SafeOpt. (Source: Sui et al. 2018)

in x is given by the upper confidence bound of the "current" GP (which is trained, contrary to the "temporary" one, only on the observations performed so far). If the lower bound of all the "temporary" GPs, computed at x is greater than the associated safety threshold, then x results in a previously unsafe point which can be considered safe with respect to the optimistic expansion from x'. Safe points x' that provide an optimistic expansion are considered as expanders.

We must remark that nobody, at our knowledge, has considered the safe optimization problem subject to both the "usual" constraints $c_i(x)$ and the safety constraints $g_i(x)$. The main difficulty in this challenge is that evaluating a point that is unsafe - in SafeOpt - implies the termination of the optimization process while evaluating an infeasible point - in CGO - only requires to update the estimate of the feasible region.

5.4 Parallel Bayesian Optimization

Performing function evaluations in parallel is a conceptually appealing way to solve optimization problems in less time that would be required by a sequential approach. Some global optimization methods, like Pure Random Search, are "embarrassingly parallel": this is not the case of BO, which is an intrinsically sequential process, so that its parallelization raises important design issues.

The main distinction between the parallel methods proposed in literature are between *synchronous* and *asynchronous* approaches. With respect to the first class of methods, a multi-point version of EI, called q-EI has been proposed in Ginsbourger et al. (2008), defined as:

$$q\mathrm{EI}\left(x_{n+1}, \ldots, x_{n+q}\right) = \mathbb{E}\left[\max\left\{\left(\min(\mathbf{y}) - y_{n+1}\right)^{+}, \ldots, \left(\min(\mathbf{y}) - y_{n+q}\right)^{+}\right\}\right]$$

where $\mathbf{y} = \{y_1, \ldots, y_n\}$ and $(\cdot)^{+}$ returns 0 if the argument is less than 0, otherwise returns the argument itself. This means to deal with a minimum of dependent random variables: the exact joint distribution of the q unknown function evaluations, $y_{n+1}, \ldots, y_{n+q}$, conditioned on previous observations through the current GP is given by:

$$\left(y_{n+1}, \ldots, y_{n+q} | D_{1:n}\right) \sim \mathcal{N}\left(\left(\mu(x_{n+1}), \ldots, \mu(x_{n+q})\right), \Sigma_q\right)$$

where Σ_q is the covariance matrix updated with the q points $x_{n+1}, \ldots, x_{n+q}$.

This synchronous parallelization approach is aimed at measuring the joint potential of a given additional set of q points to evaluate. The main idea is to consider q synchronous processors and to maximize q-EI at each iteration; then the GP model is updated with the entire set of new observations. However, as stated in Ginsbourger et al. (2008), solving the maximization of q-EI can be unaffordable when both the number of dimensions and the number of points q increase.

Computational approaches for parallelizing knowledge gradient (KG) have been also proposed (Wu and Frazier 2016). The parallel KG, namely q-KG, is Bayes-optimal for minimizing the minimum of the predictor of the GP if only one decision is remaining. The q-KG algorithm will reduce to the parallel EI algorithm in the noise-free setting and the final recommendation is restricted to the previous sampled points. The basic issue is that computing q-KG and its gradient is very expensive. Analogously to q-EI, the aim of q-KG is to identify a batch of q points to evaluate, basically in a synchronous setting. Another interesting approach, namely "portfolio allocation", has been already discussed in Chap. 4.

To address the inefficiency due to the synchronization, a new asynchronous approach, based on EI, is proposed in Ginsbourger et al. (2011). This approach, namely expectation of EI (EEI), focuses on the case of asynchronous parallel BO, where a new function evaluation is performed before all the other processors have produced their result. At every time, observations in $D_{1:n}$ are divided in three different groups: (i) already evaluated, (ii) currently under evaluation (aka "busy") and (iii) candidate points for forthcoming function evaluations. The main issue is how to take busy points into account within the GP model.

The issue on how to inject partial knowledge from an ongoing evaluation at a busy point into the EI criterion, without knowing the corresponding actual value of the function, is basically solved probabilistically:

$$\mathrm{EEI}\left(.; x^{\text{busy}}\right) = \int \mathrm{EI}\left(.; x^{\text{busy}}, y^{\text{busy}}\right)\mathrm{pdf}\left(Y\left(x^{\text{busy}}\right)|D\right)y^{\text{busy}}dy^{\text{busy}}$$

where $\mathrm{pdf}\left(Y\left(x^{\text{busy}}\right)|D\right)$ is the probability density function of the random variable $Y\left(x^{\text{busy}}\right)$ conditional on the function evaluations performed so far and stored in $D = D_{1:n}$. It is important to highlight that $EI\left(.; x^{\text{busy}}, y^{\text{busy}}\right)$ depends on y^{busy} in a quite complicated non-linear way, without any chance to have an analytical expression for EEI, even if $Y\left(x^{\text{busy}}\right)|D$ is known and simple. Therefore, EEI is approximated by averaging on a sufficient number of y^{busy} sampled according to the $Y\left(x^{\text{busy}}\right)|D$.

A straightforward way of getting statistical estimates of EEI is based on Monte Carlo sampling. The EEI approach relies on conditional simulations: it proved to be a sensible criterion, but practically not straightforward to optimize.

Finally, in Kandasamy et al. (2018) a parallel version of Thompson sampling (TS) is proposed. A theoretical analysis proves that a direct application of the sequential TS algorithm, presented in Chap. 3, in either synchronous or asynchronous parallel settings, offers a powerful result: performing n evaluations distributed among q workers is equivalent to performing n evaluations in sequence. Experimental results on simulation–optimization problems and hyperparameters optimization of a convolutional neural network prove that asynchronous TS outperforms many existing parallel BO algorithms.

5.5 Multi-objective Bayesian Optimization

Multi-objective optimization is characterized by a set of objective functions to be minimized $f_i : X \to R, i = 1, .., m$. The solution to this kind of problem is not unique and defined as $x \in X : \nexists \bar{x} \in X$, such that $f(\bar{x}) \prec f(x)$, where symbol $\prec$ denotes the dominance relation, that is:

$\forall i f_i(x) \leq f_i(\bar{x})$ and $\exists j$ s.t. $f_j(x) < f_j(\bar{x})$

This allows to identify the Pareto front, defined by the values of different objective functions and the set of corresponding dominant solutions $x \in X$.

In BO, a typical approach is to model every f_i through a Gaussian process, GP_i, with $i = 1 \ldots, m$. In Emerich and Klinkerberg (2008), the EI acquisition function has been generalized to the multi-objective case by considering the volume of the dominated region: the improvement provided by a new evaluation is the increase in the dominated volume. This extension of EI to the multi-objective case is also known as expected hyper-volume improvement (EHVI).

Given the n function evaluations performed so far, with values $f(x_i) = (f_1(x_i), \ldots, f_m(x_i))$, $i = 1, \ldots, n$, we define the set $H_n = \{y \in \mathbb{B}, \exists i \leq n, f(x_i) \prec y\}$, where $\mathbb{B} \subset \mathbb{R}^m$ is $\mathbb{B} = \{y \in \mathbb{R}^m; y \leq \bar{y}\}$ with $\bar{y} \in \mathbb{R}^m$ is introduced to guarantee that H_n is finite. Thus, H_n is the subset of $\mathbb{B}$ whose points are dominated by the previous evaluations.

Finally, the improvement provided by a new evaluation can be measured as: $I(x_{n+1}) = |H_{n+1}| - |H_n|$, where $|\cdot|$ denotes the volume in $\mathbb{R}^m$.

Since $H_{n+1} \supset H_n$, the value $I(x_{n+1})$ is the increase in the volume of the dominated region (Fig. 5.4).

Finally, maximizing EHVI means maximizing the expected improvement on $I\left(x^{n+1}\right)$ and it can be analytically defined as follows:

$$\mathrm{EHVI}(x) = \int\limits_{\mathbb{B}\setminus H_n} \mathbb{P}_n(\xi(x) \prec y)dy$$

where $\mathbb{P}_n$ is the probability conditioned on x_1, $\xi(x_1)$, ..., x_n, $\xi(x_n)$, with $\xi = (\xi_1, \ldots, \xi_m)$ is the vector of GPs modelling the corresponding objective functions $f = (f_1, \ldots, f_m)$.

In Feliot et al. (2017), the EHVI has been extended to the constrained multi-objective optimization case and to the case in which the user expresses preference-order constraints (Abdolshah et al. 2019) The EHVI acquisition function is now

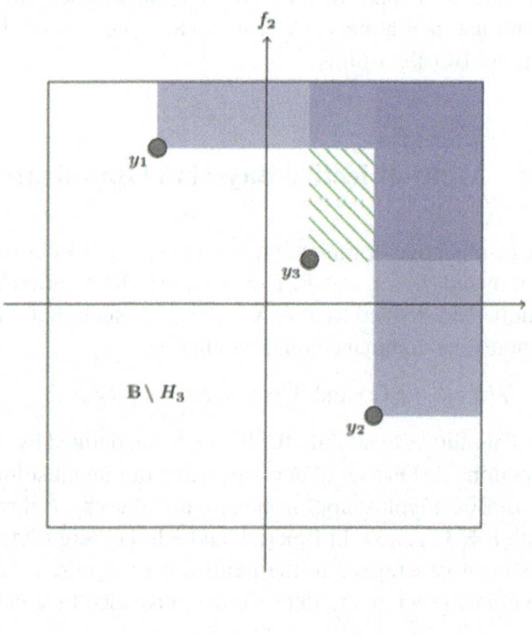

Fig. 5.4 Example of an improvement of the dominated region. The regions dominated by $\mathbf{y}_1$ and $\mathbf{y}_2$ are represented in shaded areas, with darker shades indicating overlapping regions. The hatched area corresponds to the improvement of the dominated region resulting from the observation of $\mathbf{y}_3$. (*Source* Feliot 2017)

considered as the dominant approach and some implementations are available in Matlab and the R packages "Gpareto" and "mlrMBO" (Horn 2015).

The EHVI acquisition function has been also extended in Wada and Hinu (2019) to deal with multi-point search.

An interesting application of multi-objective optimization to BO has been proposed in Calvin and Zilinskas (2019), where the acquisition of the next point to evaluate is defined as a bi-objective optimization problem. In this paper, it is shown that two well-known acquisition function, EI and PI, can be framed into the bi-objective approach, whose the objectives are $\mu(x)$ and $\sigma(x)$ of a probabilistic surrogate model.

5.6 Multi-source and Multi-fidelity Bayesian Optimization

Function evaluations in simulation-based experiments or black-box optimization are usually expensive. The basic idea in multi-fidelity is that, in the initial stages of the optimization, "cheaper" evaluation of the objective function might yield anyway considerable information while reducing the computational cost. In this framework, one is considering a collection of information sources, denoted by $f(x, s)$, where s represents a specific level of "fidelity" of the source. Usually, the lower s the higher the fidelity, with $f(x, 0)$ corresponding to the original objective function (Frazier 2018). Decreasing the fidelity decreases the accuracy of the approximation on $f(x)$, leading to a reduction in the cost of evaluation but to a poorer accuracy in the estimate of the objective function.

In multi-fidelity, the aim is $\min\limits_{x \in X} f(x)$ but observing $f(x, s)$ at a sequence of points and fidelities $\{(x_i, y_i)\}_{1:n}$ with a total cost lower than a given budget, that is $\sum_{i=1}^{n} c(s_n) \leq B$, where $c(s_n)$ is the cost to evaluate at fidelity s_n and B is the given available budget.

Acquisition functions may present some limitations in the development of multi-fidelity BO. Indeed, while KG, ES and PES can be applied, as shown in Poloczek (2017), EI is useless because evaluating $f(x, s)$ for $s \neq 0$ never offers an improvement in the best seen (i.e. EI $= 0$ for $s \neq 0$), leading to the selection of the highest fidelity source only (Frazier 2018).

Multi-fidelity optimization has been also considered for the optimization of machine learning algorithms (Klein et al. 2017; Kandasamy et al. 2019). The problem of hyperparameter tuning is also considered in Sen et al. (2018) which proposes to train the algorithm on a subsampled version of the whole dataset.

Multi-fidelity optimization has also been gaining a growing importance in several applications also in analogue circuit synthesis (Zhang et al. 2019) and engineering design (Linz et al. 2017) who propose an optimization algorithm that guides the search for solutions on a high-fidelity model through the approximation of a level set from a low-fidelity model. Using the probabilistic branch-and-bound algorithm to approximate a level set for the low-fidelity model, they are able to efficiently locate solutions inside of a target quantile, and therefore reduce the number of high-fidelity evaluations needed in searches.

An interesting application of multi-fidelity BO is presented in Perdikaris et al. (2016) for model inversion in hemodynamics and biomedical engineering (Costabal et al. 2019). A comprehensive multi-fidelity framework, in the context of bandit problems, has been proposed in Kandasamy et al. (2017) where the objective function and its approximations are sampled from a GP. Another interesting work is reported in Ghoreishi and Allaire (2018) which proposes a multi-source information approach for the constrained case.

References

Abdolshah, M., Shilton, A., Rana, S., Gupta, S., Venkatesh, S.: Multi-objective Bayesian optimisation with preferences over objectives (2019). arXiv preprint arXiv:1902.04228

Bachoc, F., Helbert, C., Picheny, V.: Gaussian process optimization with failures: classification and convergence proof (2019)

Basudhar, A., Dribusch, C., Lacaze, S., Missoum, S.: Constrained efficient global optimization with support vector machines. Struct. Multidisciplinary Optim. **46**, 201–221 (2012). https://doi.org/10.1007/s00158-011-0745-5

Berkenkamp, F., Krause, A., Schoellig, A.P.: Bayesian optimization with safety constraints: safe and automatic parameter tuning in robotics (2016). arXiv preprint arXiv:1602.04450. doi:10.1177

Calvin, J.M., Žilinskas, A.: On efficiency of bicriteria optimization. In: AIP Conference Proceedings, vol. 2070, no. 1, p. 020035. AIP Publishing (2019, February)

Candelieri, A., Archetti, F.: Sequential model-based optimization with black-box constraints: feasibility determination via machine learning. In: AIP Conference Proceedings. p. 020010 (2019)

Candelieri, A., Galuzzi, B.G., Giordani, I., Perego, R., Archetti, F.: Optimizing partially defined black-box functions under unknown constraints via Sequential Model Based Optimization: an application to Pump Scheduling Optimization in Water Distribution Networks. To appear in Proceedings of Learning and Intelligent Optimization conference (LION 13) (2019)

Candelieri, A., Giordani, I., Archetti, F., Barkalov, K., Meyerov, I., Polovinkin, A., Sysoyev, A., Zolotykh, N.: Tuning hyperparameters of a SVM-based water demand forecasting system through parallel global optimization. Comput Oper Res (2018). https://doi.org/10.1016/j.cor.2018.01.013

Costabal, F.S., Perdikaris, P., Kuhl, E., Hurtado, D.E.: Multi-fidelity classification using Gaussian processes: accelerating the prediction of large-scale computational models (2019). arXiv preprint arXiv:1905.03406

Digabel, S.L., Wild, S.M.: A taxonomy of constraints in simulation-based optimization (2015). arXiv preprint arXiv:1505.07881

Emmerich, M., Klinkenberg, J.W.: The computation of the expected improvement in dominated hypervolume of Pareto front approximations. Rapport technique, Leiden University, 34, 7–3 (2008)

Feliot, P., Bect, J., Vazquez, E.: A Bayesian approach to constrained single-and multi-objective optimization. J. Global Optim. **67**(1–2), 97–133 (2017)

Frazier, P.I.: Bayesian optimization. In: Recent Advances in Optimization and Modeling of Contemporary Problems, pp. 255–278. INFORMS (2018)

Gardner, J.R., Kusner, M.J., Xu, Z.E., Weinberger, K.Q., Cunningham, J.P.: Bayesian Optimization with Inequality Constraints Jacob. In: ICML, pp. 937–945 (2014)

Ghoreishi, S.F., Allaire, D.: Multi-information source constrained Bayesian optimization. Struct. Multidisciplinary Optim. 1–15 (2018). https://doi.org/10.1007/s00158-018-2115-z

Ginsbourger, D., Riche, R. Le, Carraro, L.: A multi-points criterion for deterministic parallel global optimization based on Gaussian processes. In: International Conference on Nonconvex Programming, NCP07, Rouen, France. 1–30 (2008)

Ginsbourger, D., Riche, R. Le: Dealing with asynchronicity in parallel Gaussian Process based global optimization $*$. In: 4th International Conference of the ERCIM WG on Computing & Statistics (ERCIM'11) (2011)

Gramacy, R.B., Lee, H.K.H.: Optimization under unknown constraints. Bayesian Statistics, 9 (2011)

Gramacy, R.B., Gray, G.A., Le Digabel, S., Lee, H.K., Ranjan, P., Wells, G., Wild, S.M.: Modeling an augmented Lagrangian for blackbox constrained optimization. Technometrics **58**(1), 1–11 (2016)

Hernández-Lobato, J.M., Gelbart, M.A., Hoffman, M.W., Adams, R.P., Ghahramani, Z.: Predictive entropy search for bayesian optimization with unknown constraints (2015)

Horn, D., Wagner, T., Biermann, D., Weihs, C., Bischl, B.: Model-based multi-objective optimization: Taxonomy, multi-point proposal, toolbox and benchmark. In: International Conference on Evolutionary Multi-Criterion Optimization, pp. 64–78. Springer, Cham (2015, March)

Jones, D.R., Schonlau, M., Welch, W.J.: Efficient global optimization of expensive black-box functions. J. Glob. Optim. **13**(4), 455–492 (1998)

Kandasamy, K., Dasarathy, G., Schneider, J., Póczos, B.: Multi-fidelity bayesian optimisation with continuous approximations. In: Proceedings of the 34th International Conference on Machine Learning, vol. 70, pp. 1799–1808. JMLR. org (2017, August)

Kandasamy, K., Krishnamurthy, A., Schneider, J., Poczos, B.: Asynchronous parallel Bayesian optimisation via Thompson sampling (2018). arXiv preprint arXiv:1705.09236

Kandasamy, K., Neiswanger, W., Schneider, J., Poczos, B., Xing, E.: Neural architecture search with Bayesian optimisation and optimal transport (2019). arXiv preprint arXiv:1802.07191

Klein, A., Falkner, S., Bartels, S., Hennig, P., Hutter, F.: Fast Bayesian optimization of machine learning hyperparameters on large datasets.In: Proceedings of the 20th International Conference on Artificial Intelligence and Statistics. Proceedings of Machine Learning Research, Brookline, MA, 54, 528–536 (2017)

Letham, B., Karrer, B., Ottoni, G., Bakshy, E.: Constrained Bayesian optimization with noisy experiments. Bayesian Anal. **14**(2), 495–519 (2019)

Linz, D.D., Huang, H., Zabinsky, Z.B.: Multi-fidelity simulation optimization with level set approximation using probabilistic branch and bound. In: Winter Simulation Conference (WSC), IEEE, pp. 2057–2068 (2017, December)

Perdikaris, P., Karniadakis, G.E.: Model inversion via multi-fidelity Bayesian optimization: a new paradigm for parameter estimation in haemodynamics, and beyond. J. R. Soc. Interface **13**(118), 20151107 (2016)

Picheny, V., Gramacy, R. B., Wild, S., Le Digabel, S.: Bayesian optimization under mixed constraints with a slack-variable augmented Lagrangian. In: Advances in Neural Information Processing Systems, pp. 1435–1443 (2016)

Poloczek, M., Wang, J., Frazier, P. Multi-information source optimization. In: Guyon, I., Luxburg, U.V., Bengio, S., Wallach, H., Fergus, R., Vishwanathan, S., Garnett, R. (eds.) Advances in Neural Information Processing Systems, 30, pp. 4291–4301. Curran Associates, Red Hook, NY, (2017)

Rudenko, L.I.: Objective functional approximation in a partially defined optimization problem. J. Math. Sci. **72**(5), 3359–3363 (1994)

Sacher, M., Duvigneau, R., Le Maitre, O., Durand, M., Berrini, E., Hauville, F., Astolfi, J.A.: A classification approach to efficient global optimization in presence of non-computable domains. Struct. Multidisciplinary Optim. **58**(4), 1537–1557 (2018)

Sen, R., Kandasamy, K., Shakkottai, S.: Multi-fidelity black-box optimization with hierarchical partitions. In: International Conference on Machine Learning, pp. 4545–4554 (2018, July)

Sergeyev, Y.D., Kvasov, D.E.: Deterministic Global Optimization: an Introduction to the Di-agonal Approach. Springer (2017)

Sergeyev, Y.D., Kvasov, D.E., Khalaf, F.M.: A one-dimensional local tuning algorithm for solving GO problems with partially defined constraints. Optim. Lett. **1**(1), 85–99 (2007)

Sui, Y., Gotovos, A., Burdick, J.W., Krause, A.: Safe exploration for optimization with gaussian processes. In: International Conference on Machine Learning (ICML), (SafeOpt) (2015)

Sui, Y., Zhuang, V., Burdick, J.W., Yue, Y.: Stagewise safe Bayesian optimization with Gaussian processes (2018). arXiv preprint arXiv:1806.07555

Tsai, Y.A., Pedrielli, G., Mathesen, L., Zabinsky, Z.B., Huang, H., Candelieri, A., Perego, R.: Stochastic optimization for feasibility determination: an application to water pump operation in water distribution network. In: Proceedings of the 2018 Winter Simulation Conference, pp. 1945–1956. IEEE Press (2018, December)

Vapnik, V.: Statistical Learning Theory. Willey, New York (1998)

Wada, T., Hino, H.: Bayesian optimization for multi-objective optimization and multi-point search (2019). arXiv preprint arXiv:1905.02370

Wu, J., Frazier, P.: The parallel knowledge gradient method for batch bayesian optimization. In: Advances in Neural Information Processing Systems, pp. 3126–3134 (2016)

Zhang, S., Lyu, W., Yang, F., Yan, C., Zhou, D., Zeng, X., Hu, X.: An efficient multi-fidelity Bayesian optimization approach for analog circuit synthesis. In: DAC, pp. 64–1 (2019, June)

Chapter 6
Software Resources

This chapter is aimed to introduce frameworks developed in the Bayesian optimization (BO) community describing many aspects of these tools, where it is possible to retrieve the tool, in which language and version are available. Section 6.1 is devoted to open source software, while industrial BO as a service solutions are analysed in 6.2. Section 6.3 is devoted to BO solutions specifically developed for hyperparameters optimization of machine learning algorithms. Section 6.4 presents relevant sources for test problems, both in terms of test functions and generators. Finally, Sect. 6.5 reports some relevant non-Bayesian global optimization software. The software in this section refers, basically, to the box-constrained case, with the exception of Predictive Entropy Search with Constraints (PESC) which is included in the open source package Spearmint (https://github.com/HIPS/Spearmint/tree/PESC).

6.1 Open Source Software

In BO community, there are currently a variety of frameworks that implement the sequential optimization process through Bayesian optimization. Several of them use different combinations between the surrogate model and the acquisition function. However, most of them adopt Gaussian process (GP) as a surrogate model inside of optimization process, with several acquisition functions as probability of improvement (PI), expected improvement (EI), upper-confidence bound (UCB), lower-confidence bound (LCB) and others.

mlrMBO (https://github.com/mlr-org/mlrMBO) (Bischl et al. 2017) is a flexible and comprehensive R toolbox for BO. It is designed for both single and multi-objective optimization with mixed continuous, categorical and conditional parameters.

More advanced features are implemented such as parallelization, multi-point batch proposal and visualization of optimization improving.

mlrMBO is designed in a modular style, so each component can easily be replaced or adapted to specific use cases. Indeed, it is possible to use the main regression

F. Archetti and A. Candelieri, *Bayesian Optimization and Data Science*,
SpringerBriefs in Optimization, https://doi.org/10.1007/978-3-030-24494-1_6

models inside the machine learning library R named *mlr*, such as GP, Random Forest (RF). It is also possible to define from scratch a new acquisition function and change the optimizer.

Spearmint (https://github.com/HIPS/Spearmint), with an older version under a different licence (available at https://github.com/JasperSnoek/spearmint), is a Bayesian optimization library written in Python 2.7. Spearmint is the result of a collaboration primarily between machine learning researchers at Harvard University and the University of Toronto. It was implemented by the founders of the Bayesian optimization start-up *Whetlab*, which was acquired by Twitter in 2015. This tool uses as a surrogate model only GP and adopts EI as an acquisition function.

Metrics Optimization Engine (***MOE***) (https://github.com/Yelp/MOE) is a Bayesian optimization library written in C++ with a Python 2.7 wrapper that supports GPU-based computations for improved speed. It was developed at Yelp and supports only GP surrogate model with just EI as an acquisition function. Another version *Cornell MOE* (https://github.com/wujian16/Cornell-MOE), developed at Cornell University, is built on MOE, and it is easier to install and support parallel optimization and optimization with derivatives. Indeed, this new version has several new acquisition functions that take into account the derivative information. In addition to EI acquisition function, derivative information EI (d-EI), batch knowledge gradient (d-KG) and without derivative information knowledge gradient (q-KG) are implemented.

GPyOpt (https://github.com/SheffieldML/GPyOpt) is a Bayesian optimization library written in Python, for both versions 2 and 3, implemented and maintained by the machine learning group at Sheffield University. The initial package was based on top of the Gaussian process regression library GPy (https://sheffieldml.github.io/GPy/) written and maintained for the same group of BO tool. Currently, the surrogate models available in this package are GP, warped GP and RF. The main acquisition functions implemented are PI, EI, LCB and Entropy Search (ES).

pyGPGO (https://github.com/hawk31/pyGPGO) (Jiménez and Ginebra 2017) is a flexible BO library written in Python > 3.5 and implemented in a modular fashion. The software was developed by the collaboration between Universitat Pompeu Fabra and Universitat Politècnica de Catalunya. With this library, it is possible to use several surrogate models that include GP, RF, BoostedTrees (BT), t-Student Process (tSP) and t-Student Process MCMC (tSPMCMC). It is also possible to define and add a novel surrogate model for this library. The acquisition functions available are PI, EI, UCB and ES.

pyBO (https://github.com/mwhoffman/pybo) (Hoffman and Shahriari 2014) is a BO library written in Python, for both version 2 and 3, by Matthew W. Hoffman and Bobak Shahriari (https://github.com/mwhoffman/pybo). It uses only a GP as a surrogate model and implements several acquisition functions, which are PI, EI, UCB, Thompson sampling (TS).

*Sequential **M**odel-based **A**lgorithm **C**onfiguration* (SMAC) (http://www.cs.ubc.ca/labs/beta/Projects/SMAC/) (Hutter et al. 2011) is a library for optimizing algorithm hyperparameters written Java. Currently, the original version was re-implemented in Python 3 with name *SMAC3* (https://github.com/automl/SMAC3). This library is developed by the AutoML Group of the University of Freiburg (https://www.automl.org/automated-algorithm-design/algorithm-configuration/smac/). For the surrogate model, this library includes GP (from *George* package), RF (from *pyrfr* package) and EI, PI and LCB for acquisition functions.

Robust **B**ayesian **O**ptimization (RoBo) (https://github.com/automl/RoBO) (Klein et al. 2017) is a library written in Python 3. This library is implemented and maintained form Machine Learning Lab from the University of Freiburg (https://ml.informatik.uni-freiburg.de/). Its structure is a modular and that allows to easily add and exchange components in Bayesian optimization such as different surrogate models and acquisition functions. For the surrogate model, this library includes GP (from *George* package), RF (from *pyrfr* package) and a native Bayesian neural networks implementation. Also, the software for acquisition function includes EI, PI, LCB and information gain. (https://www.automl.org/automl/robo/).

Random **Em**beddings **B**ayesian **O**ptimization (REMBO) (https://github.com/ziyuw/rembo) is a MATLAB package for applying Bayesian optimization to highly dimensional problems. This library offers only GP as a surrogate model and UCB as acquisition function.

Scikit-Optimize (https://scikit-optimize.github.io/) is a modular BO library written in Python, for both versions 2 and 3, by the same authors of the most famous Python machine learning library named Scikit-Learn (https://scikit-learn.org/stable/). Based on the same prefix on its name, each surrogate model that can be used inside of this package is wrapped on Scikit-Learn library, so is possible to use most common surrogate models including GP, RF and Gradient Boosted Regression Trees (GBRT).

fmfn/BayesianOptimization (https://github.com/fmfn/BayesianOptimization) is a BO library written in Python, for both version 2 and 3. This package uses only GP as a surrogate model using the Scikit-Learn library (https://scikit-learn.org/stable/) (Pedregosa et al. 2011) and natively implemented inside the most common acquisition functions that are PI, EI and UCB. A simple tutorial (http://philipperemy.github.io/visualization/) is provided to show how to use this library.

GPflowOpt (https://github.com/GPflow/GPflowOpt) (Knudde et al. 2017) is a Python package for BO design goals focus on a framework that can be extended with custom acquisition functions and models. GPflowOpt includes some standard acquisition functions as PI, EI, LCB. Besides, also implements the max-value entropy search (MES) (Wang and Jegelka 2017). Finally for BO surrogate model, this framework uses GPflow (https://github.com/GPflow/GPflow) Python library for the GP implementations, based on using the TensorFlow library.

Auto-Keras (https://github.com/jhfjhfj1/autokeras/https://autokeras.com/) is an open source Python 3.6 library for automated machine learning (AutoML). It is developed by DATA Lab at Texas A&M University and community contributors. Auto-Keras provides functions to automatically search for architecture and hyperparameters of deep learning models with Keras framework.

Hyperopt (https://github.com/hyperopt/hyperopt) (Bergstra et al. 2015) is a Python library for serial and parallel optimization over awkward search spaces, which may include real-valued discrete and conditional dimensions. Currently, two search methods are implemented in this library: Random Search and Tree Parzen Estimator (TPE). In addition, there is a version of this package, named *Hyperopt-sklearn* (https://github.com/hyperopt/hyperopt-sklearn) specialized to optimize the hyperparameters of models of Scikit-Learn machine learning library. In the future, GP and RF will be implemented as surrogate models. A simple tutorial (https://towardsdatascience.com/an-introductory-example-of-bayesian-optimization-in-python-with-hyperopt-aae40fff4ff0) is provided to show how to use this library.

bayesopt (https://it.mathworks.com/help/stats/bayesian-optimization-workflow.html#bva8nmd-1) is a MATLAB function available only inside *Statistics and Machine Learning Toolbox* in MATLAB environment. Primarily target to optimize regression and classification models, it can also be used to optimize different objective functions. For the surrogate model, the package provides just GP, while there are some acquisition functions, such as EI, expected improvement plus, LCB and PI.

BayesOpt (https://github.com/rmcantin/bayesopt—https://bitbucket.org/rmcantin/bayesopt/) (Martinez-Cantin 2014) is a library written in C++ extremely efficient, portable and flexible. It offers common interfaces for several programming languages C, C++, Python, MATLAB and Octave. The surrogate model, available in this package, is only GP, and there are PI, EI, LCB and TS as acquisition functions.

PARyOpt *Parallel Asynchronous Remote Bayesian Optimization* (***PARyOpt***) (https://bitbucket.org/baskargroup/paryopt) (Pokuri et al. 2018) is a Python 3.5 package for Bayesian optimization that also implements a parallel asynchronous version of this efficient global optimization method. Besides, with this library is possible to run the parallel optimization process in local, in remote or in their combination. This package implements only GP as a surrogate model, and it utilizes the common acquisition functions as PI, EI, LCB and UCB.

Pyro (https://github.com/pyro-ppl/pyro) (Bingham et al. 2019) is a probabilistic programming language (PPL) developed at Uber AI Labs for implementing probabilistic models. It is a library written in Python 3 that can perform Bayesian optimization using GP as a surrogate model, and it is possible to implement a custom acquisition function.

ProBO (https://github.com/willieneis/ProBO) (Neiswanger et al. 2019) is another software environment based on probabilistic programming specific for BO. Probabilistic programs are model tools that allow flexible model composition incorporating prior information and enabling automatic inference. Beyond the usual GP, ProBO allows to choose from a library including latent factor models, deep Bayesian networks, hierarchical regression models and used them in BO.

G3PO (Github missing) (Lavin 2018) is an open source library written in C++ that uses a combination of Bayesian optimization and probabilistic programming. Inside of this library is available only as a modified GP, as a surrogate model, that allows embedding of prior information and only EI as acquisition function.

Probabilistic Harvard Optimizer Exploring Non-Intuitive Complex Surfaces (***PHOENICS***) (https://github.com/aspuru-guzik-group/phoenics) (Häse et al. 2018) is an open source Python 3.6 library that combines Bayesian optimization using Bayesian kernel density estimation (BKDE) as a surrogate model.

Sherpa (https://github.com/sherpa-ai/sherpa) (Hertel et al. 2018) is an open source Python 3 library for hyperparameters optimization using a variety of global optimization methods such as Random Search, Grid Search, Local Search, Bayesian Optimization and Population-Based Training (PBT). In specific for Bayesian optimization process, this library wraps the unique surrogate model GP from GPyOpt library and uses only EI as an acquisition function. This library has a graphical interface for seeing the improvements during the optimization process. The latest documentation is available here (https://parameter-sherpa.readthedocs.io/en/latest/), and there are also available some useful tutorials of using this library (https://www.youtube.com/watch?v=L95sasMLgP4&feature=youtu.be).

Fast LineAr SearcH (***FLASH***) (Nair et al. 2018; Zhang et al. 2016) is an open source Python 2.7 and 3 library for AutoML. In specific, this package performs a two-layer Bayesian optimization, which first uses a parametric model to select the most promising algorithms, then computes a nonparametric model to fine-tune hyperparameters of the promising algorithms on a variety of dataset. The library has GP and RF as surrogate models and implements EI and expected improvement per second (EIPS) (Snoek et al. 2012) as acquisition functions.

COMmon Bayesian Optimization Library (***COMBO***) (https://github.com/tsudalab/combo) is an open source Python 2.7 library for global optimization. This package allows optimizing problem with high dimensionality by Bayesian optimization method. It implements EI, PI and TS as acquisition functions and GP as a surrogate model.

Dragonfly (https://github.com/dragonfly/dragonfly) (Kandasamy et al. 2019) is an open source Python (2.7 and 3.5) library for scalable and robust BO specific for hyperparameters optimization. This library allows optimizing efficiently high-dimensional problems exploiting the additive structure of the objective function, and, adopts multi-fidelity optimization and parallelization techniques. Currently, GP is the unique surrogate model implemented, and the following acquisition functions are available in this library: PI, EI, UCB, top-two expected improvement (TTEI), TS, Add-GP-UCB. Also, a comparison between open source Bayesian optimization implementations is available on this tutorial: http://ash-aldujaili.github.io/blog/2018/04/01/coco-bayesopt/.

6.2 Bayesian Optimization as a Service

In addition to the libraries introduced previously, there are other libraries, always related to the Bayesian optimization paradigm, which allows optimizing the particular objective function through the connection to a remote optimization service using a cloud distributed computational power. With these libraries, the user, who has no advanced knowledge in optimization methods, can adopt a single or a combination of very sophisticated search methods. Indeed, for the majority of the following software tools, the user is blind to the methods and architecture of each optimization system and uses it through an API.

SigOpt (http://sigopt.com) is an example of "Black-Box Optimization as a Service". The core of this service was initially created for a project named MOE by the

Cornell University. The user has to interact through REST API in several languages like Java, R and Python with the most updated version of each language. The main idea behind the optimization service is that the user is totally blind to the computational architecture infrastructure which uses with Bayesian optimization process. To use this service, the user has to subscribe an Academia or Enterprise account.

OPTaaS is another black-box optimization service accessible via API in R and Python languages, published by the company Mind Foundry. It deals with any type of parameters (e.g. continuous, discrete, categorical, conditional). To use this service, the user has to subscribe an Academia or Commercial account.

A simple tutorial (https://towardsdatascience.com/how-to-visualize-black-box-optimization-problems-with-gaussian-processes-a6128f99b09d) is provided to show how to use this optimization service.

VUKU (https://www.prowler.io/) is another example of "Black-Box Optimization as a Service". This service is produced by *PROWLER.io* company. The structure of this optimization service is more complex respect to above systems because uses in the same optimization process multiple decision models.

SVM-CBO, presented in Chap. 5, is provided as a service by OAKS Optimization Analytics Knowledge and Simulation (OAKS) (http://www.oaks.cloud/). SVM-CBO addresses black-box optimization in presence of simulation optimization, partially defined functions and non-computable domains.

6.3 Bayesian Optimization-Based Services for Hyperparameters Optimization

SAS-Autotune (Koch et al. 2018) is a framework implemented in a particular SAS language named PROC CAS (Cloud Analytic Services—https://documentation.sas.com/?docsetId=proccas&docsetTarget=p09ouj759e7ysyn1b0ws8f0ho9e5.htm&docsetVersion=3.1&locale=en) which is based on common algebraic specification language (CASL). The system is designed to perform optimization of general nonlinear functions over both continuous and integer variables, but it is mainly specialized in optimizing hyperparameters in machine learning models. The system can be run in a single machine mode or in distributed mode. The problem to optimize could be single- or multi-objective.

This system combines a number of specialized sampling and search methods that are very effective in tuning machine learning models. The search method used by SAS-Autotune is genetic algorithm (GA), generating set search (GSS), Bayesian optimization, DIRECT, Nelder-Mead and DIRECT hybrid.

Amazon SageMaker Automatic Model Tuning (https://aws.amazon.com/it/blogs/aws/sagemaker-automatic-model-tuning/) is a service offered by Amazon for its machine learning platform Amazon SageMaker. The user can use this service through SageMaker Python SDK, which gives to the user the API access to service of hyperparameters tuner on Amazon cloud platform.

Google Vizier (https://ai.google/research/pubs/pub46180) (Golovin et al. 2017) is a Google internal service for performing black-box optimization that has become the parameter tuning of the engine at Google Cloud Platform (GCP). In specific, this framework is used to optimize many of Google machine learning models and other systems. Also, it constitutes the main capabilities core of *HyperTune*, a hyper-parameter tuner in Google Cloud Machine Learning Engine on GCP. The system is available in a "blind way", as a service, to users that use Machine Learning Engine on GCP, currently is possible to have a local open source version of Google Vizier on GitHub written in Python 2.7 (https://github.com/tobegit3hub/advisor).

Internally, this cloud service or GitHub library version is able to optimize an objective function with several methods including the most common methods: Grid Search, *Bayesian optimization*, Random Search and Metaheuristic approaches as Simulate Anneal, Covariance Matrix Adaptation Evolution Strategy (CMAES) and Multi-objective Covariance Matrix Adaptation Evolution Strategy (MOCMAES) (Tables 6.1 and 6.2).

6.4 Test Functions and Generators

In this section, we will introduce some benchmark functions (with surveys and sites/repositories) reported in the literature of optimization community and generators for test functions that are used to evaluate the capacity of the optimization framework/tool in terms of efficacy and effectiveness.

6.4.1 Survey and Site/Repository of Test Functions

In Jamil and Yang (2013), a rich set of 175 benchmark functions are described for unconstrained optimization problems with diverse properties in terms of modality, separability, and valley landscape. Besides, in Suganthan et al.(2005) are reported 28 benchmark functions divided into three categories: unimodal functions, multimodal function and composition functions.

For each test function, their implementations are available for R (https://cran.r-project.org/web/packages/cec2013/cec2013.pdf), MATLAB, C and Java (http://web.mysites.ntu.edu.sg/epnsugan/PublicSite/Shared Documents/Forms/AllItems.aspx) languages.

Virtual Library of Simulation Experiments: Test functions and datasets are a site/repository (https://www.sfu.ca/~ssurjano/optimization.html) that reports a suite of functions and datasets for evaluating new approaches to the design and analysis of experiments involving computer models. In specific, there is a section inside of this repository dedicated only for the optimization problem, where benchmark functions are grouped according to similarities in their significant physical properties and shapes. For each test function a detailed description is provided as well as implemen-

Table 6.1 Open source BO tools

Name tool	Surrogate implementation	Surrogates available	Acquisition functions	Licence	Last update	Language version
mlrMBO	mlr	{GP, RF}	{EI, CB, AEI, EQI, DIB, AdaCB}	BSD-2	06/2018	R > = 3.5.1
Spearmint	Native	{GP}	{EI}	Academic	12/2016	Python 2.7
MOE	Native	{GP}	{EI}	Apache	04/2016	Python 2.7
GPyOpt	GPy	{GP, RF, WGP}	{PI, EI, UCB}	BSD-3	04/2017	Python 2/3
pyGPGO	Native (GP, Tsp, GBM), Scikit-Learn (RF, BT)	{GP, RF, BT, Tsp, GBM}	{PI, EI, UCB, Entropy}	MIT	01/2019	Python > = 3.5
pyBO	Reggie	{GP}	{PI, EI, UCB, Thompson sampling}	BSD-2	09/2015	Python 2/3
fmfn/Bayesian Optimization	Scikit-Learn	{GP}	{PI, EI, UCB}	MIT	03/2017	Python 2/3
SMAC	George (GP), pyrfr (RF)	{GP, RF}	{PI, EI, LCB}	AGPLv3	01/2019	Python 3, Java
RoBO	George (GP), pyrfr (RF)	{GP, RF, BNN}	{PI, EI, LCB, information gain}	BSD-3	02/2019	Python 3
REMBO	Native	{GP}	{EI, UCB}	?	08/2013	MATLAB
Auto-Keras	Native	{GP}	{UCB}	MIT	02/0219	Python 3.6
Hyperopt	Native	{TPE, SA, Random Search}	–	BSD-3	02/2019	Python 3
GpflowOpt	Gpflow	{GP}	{PI, EI, LCB, MES}	Apache 2.0	10/2018	Python 3.6
Scikit-Optimize	Scikit-Learn	{GP, RF, GBRT}	{PI, EI, UCB}	BSD	04/2017	Python 2/3
bayesopt	Native	{GP}	{PI, EI, LCB, EI Plus}	?	02/2019	MATLAB

(continued)

Table 6.1 (continued)

Name tool	Surrogate implementation	Surrogates available	Acquisition functions	Licence	Last update	Language version
BayesOpt	Native	{GP}	{PI, EI, LCB, TS}	AGPL	09/2018	C++ (C, Python, MATLAB, Octave)
PARyOpt	Native	{GP}	{PI, EI, LCB, UCB}	MIT	02/2019	Python 3.5
Pyro				MIT		Python 2/3
ProBO	GPy	{GP}	{PI, EI, UCB, TS}	–	–	–
G3PO	?	{GP}	{EI}	?	?	C++
Phoenics	Native	{BKDE}	–	Apache 2.0	01/2019	Python 3.6
Sherpa	GPyOpt	{GP}	{EI}	GPLv3	02/2019	Python 3
FLASH	?	{GP, RF}	{EI, EIPS}	GPLv3	09/2017	Python 2
COMBO	Native	{GP}	{PI, EI, TS}	MIT	04/2016	Python 2/3
Dragonfly	Native	{GP}	{PI, EI, UCB, TTEI, TS, Add-GP-UCB}	MIT	03/2019	Python 2.7/3.5

Table 6.2 Industrial BO tools

Name tool	Licence	Language version
Sigopt	Commercial/academic	R, Python, Java
Google Vizier	Commercial	(Python 2.7 for open source version)
OPTaaS	Commercial/academic	R, Python
SAS-Autotune	Commercial	PROC CAS
Amazon SageMaker Automatic Model Tuning	Commercial	Python
VUKU	Commercial/academic(?)	–

tations in both MATLAB and R. The website is in continuous updating and more functions are being added in the next future.

Global Optimization: Methods and codes are a site/repository (http://www-optima.amp.i.kyoto-u.ac.jp/member/student/hedar/Hedar_files/TestGO.htm) that reports a suite of test functions for unconstrained and constrained global optimization. There are detailed descriptions and implementation in MATLAB language of these benchmark functions. The repository maintainer is Dr. Abdel-Rahman Hedar from Computer Science and Information Sciences department of Assiut University Egypt.

globalOptTests (https://cran.rstudio.com/web/packages/globalOptTests/globalOptTests.pdf) is a R package that maps many well-known test functions for unconstrained global optimization from GAMS notation to R functions. This repository/package was proposed in 2014 in (Mullen 2014).

benchfunk (https://github.com/mwhoffman/benchfunk) is a Python 3 package that provides a set of most common test functions (e.g. Sinusoidal, Branin, Hartamann 3D/6D, Goldestein) that can be used to benchmark global optimization and Bayesian optimization algorithm.

optimization–evaluation (https://github.com/keit0222/optimization-evaluation) is a Python 3 package that provides a set of benchmark functions divided into unimodal functions, with single and multiple global minimum, and multi-modal functions with a single global minimum (https://qiita.com/tomitomi3/items/d4318bf7afbc1c835dda#ackley-function).

DEAP benchmarks (https://github.com/deap/deap) (De Rainville et al. 2012) is a novel evolutionary computation framework written in Python 3 that has inside a benchmarks test functions module (https://deap.readthedocs.io/en/master/api/benchmarks.html#). This module contains a single and multi-objective test functions implementations.

6.4.2 Test Functions Generators

GKLS (Gaviano et al. 2003) is a generator of several test functions, providing the capability to modulate the "difficulty" of the global optimization problem. The creators of this generator, from which the name derives, are **G**aviano M., **K**vasov D. E., **L**era D., **S**ergeyev Y. D. By this generator, it is possible to generate non-differentiable, continuously differentiable and twice continuously differentiable classes of test functions for multi-extremal multidimensional box-constrained global optimization. For each test class, it is possible to define 100 functions with arbitrary dimensions. Test functions are generated by defining a convex quadratic function systematically distorted by polynomials in order to introduce local minima. Besides the position of the global minimum, this generator also defines the exact local minimum positions of the test function. The parameters that user has to define for generating a test function are five as following: problem dimension, number of local minima, the value of global minimum, the radius of the attraction region of the global minimizer and the distance from the global minimizer to the vertex of the quadratic function.

Global Constrained optimization problem Generator (GCGen) (Gergel et al. 2019) is a generator of constrained test functions. With GCGen, the modified GKLS function was used as the objective function, and the constraints were the exterior of the balls with given centres and radii. For generating the test functions, in addition to the traditional GKLS inputs (e.g. number of dimensions, the global minimum value, the radius of the attraction region of the global minimum, etc.), one needs to provide the number of constraints and the feasible fractions on the function's domain. This generator is available as open source at Github. (https://github.com/UNN-ITMM-Software/GCGen).

6.5 Non-Bayesian Global Optimization Software

Since this topic is outside the focus of this book, what follows is just an indication of well-known and widely used global optimization software.

- **Lipschitz Global Optimizer (LGO)** suite can handle in principle "all" models defined by merely continuous or Lipschitz(-continuous) functions. The optimization model to solve can be handed over to LGO as "black box", without the need for explicitly given model functions. This key feature makes LGO applicable in a large variety of contexts (https://www.pinterconsulting.com/).
- **BARON** (https://minlp.com/baron) is a commercial global optimization software offering deterministic guarantee for global optimality of nonlinear and mixed-integer nonlinear problems. BARON is a general purpose solver that implements a branch-and-reduce algorithm for solving mixed-integer nonlinear optimization problems. Purely continuous, purely integer and mixed-integer nonlinear problems can be solved with the software. BARON can be used in combination with **ALAMO** (https://minlp.com/alamo) which provides a machine learning-based surrogate model.

References

Bergstra, J., Komer, B., Eliasmith, C., Yamins, D., Cox, D.D.: Hyperopt: a Python library for model selection and hyperparameter optimization. Comput. Sci. Discovery. **8** (2015). https://doi.org/10.1088/1749-4699/8/1/014008

Bingham, E., Chen, J.P., Szerlip, P., Goodman, N.D.: Pyro: deep universal probabilistic programming. J. Mach. Learn. Res. **20**, 1–6 (2019)

Bischl, B., Richter, J., Bossek, J., Horn, D., Thomas, J., Lang, M.: mlrMBO: a modular framework for model-based optimization of expensive black-box functions (2017). arXiv preprint arXiv:1703.03373. https://doi.org/10.13140/rg.2.2.11865.31849

De Rainville, F.-M., Fortin, F.-A., Gardner, M.-A., Parizeau, M., Gagné, C.: DEAP: a Python framework for evolutionary algorithms. In: Proceedings of the 14th Annual Conference Companion on Genetic and Evolutionary Computation, pp. 85–92. ACM (2012)

Gaviano, M., Kvasov, D.E., Lera, D., Sergeyev, Y.D.: Algorithm 829: software for generation of classes of test functions with known local and global minima for global optimization. ACM Trans. Math. Softw. **29**, 469–480 (2003)

Gergel, V., Barkalov, K., Lebedev, I., Rachinskaya, M., Sysoyev, A.: A flexible generator of constrained global optimization test problems. In: Proceedings LeGO—14th International Global Optimization Workshop, p. 020009 (2019)

Golovin, D., Solnik, B., Moitra, S., Kochanski, G., Karro, J., Sculley, D.: Google Vizier: A service for black-box optimization. In: Proceedings of the 23rd ACM SIGKDD International Conference on Knowledge Discovery and Data Mining, pp. 1487–1495. ACM (2017)

Häse, F., Roch, L.M., Kreisbeck, C., Aspuru-Guzik, A.: Phoenics: a Bayesian optimizer for chemistry. ACS Central Sci. **4**, 1134–1145 (2018)

Hertel, L., Sadowski, P., Collado, J., Baldi, P.: Sherpa : Hyperparameter Optimization for Machine Learning Models (2018)

Hoffman, M. W., Shahriari, B.: Modular mechanisms for bayesian optimization. In NIPS Workshop on Bayesian Optimization, pp. 1–5 (2014)

Hutter, F., Hoos, H.H., Leyton-Brown, K.: Sequential model-based optimization for general algorithm configuration. In: International Conference on Learning and Intelligent Optimization, pp. 507–523. Springer, Berlin, Heidelberg (2011, January)

Jamil, M., Yang, X.-S.: A literature survey of benchmark functions for global optimization. Int. J. Math. Model. Numer. Optim. **4**, 150–194 (2013). https://doi.org/10.1504/ijmmno.2013.055204

Jiménez, J., Ginebra, J.: pyGPGO: Bayesian optimization for Python. J. Open Source Softw. **2**, 431 (2017). https://doi.org/10.21105/joss.00431

Kandasamy, K., Vysyaraju, K.R., Neiswanger, W., Paria, B., Collins, C.R., Schneider, J., Poczos, B.. Xing, E.P.: Tuning hyperparameters without grad students: scalable and robust Bayesian optimisation with Dragonfly (2019). arXiv preprint arXiv:1903.06694

Klein, A., Falkner, S., Mansur, N., Hutter, F.: RoBO: a flexible and robust Bayesian optimization framework in Python. In: Conference on Neural Information Processing Systems (NIPS): Bayesian Optimization Workshop (2017)

Knudde, N., van der Herten, J., Dhaene, T., Couckuyt, I.: GPflowOpt: A Bayesian optimization library using tensorflow (2017). arXiv preprint arXiv:1711.03845

Koch, P., Golovidov, O., Gardner, S., Wujek, B., Griffin, J., Xu, Y.: Autotune: a derivative-free optimization framework for hyperparameter tuning. In: Proceedings of the 24th ACM SIGKDD International Conference on Knowledge Discovery & Data Mining. pp. 443–452 (2018)

Lavin, A.: Doubly Bayesian optimization (2018). arXiv preprint arXiv:1812.04562

Martinez-Cantin, R.: BayesOpt: a Bayesian optimization library for nonlinear optimization, experimental design and bandits. J. Mach. Learn. Res. **15**, 3735–3739 (2014)

Mullen, K.M.: Continuous global optimization in R. J. Stat. Softw. **60**(6), 1–45 (2014)

Nair, V., Yu, Z., Menzies, T., Siegmund, N., Apel, S.: Finding faster configurations using FLASH (2018). arXiv preprint arXiv:1801.02175

Neiswanger, W., Kandasamy, K., Poczos, B., Schneider, J., Xing, E.: ProBO: a Framework for using probabilistic programming in Bayesian optimization(2019). arXiv preprint arXiv:1901.11515

Pedregosa, F., Michel, V., Grisel, O., Blondel, M., Prettenhofer, P., Weiss, R., Vanderplas, J., Cournapeau, D.: Scikit-learn: machine learning in Python. J. Mach. Learn. Res. **12**, 2825–2830 (2011). https://doi.org/10.1007/s13398-014-0173-7.2

Pokuri, B.S.S., Lofquist, A., Risko, C.M., Ganapathysubramanian, B.: PARyOpt: a software for parallel asynchronous remote Bayesian optimization (2018). arXiv preprint arXiv:1809.04668; arXiv:1809.04668v1

Snoek, J., Larochelle, H., Adams, R.P.: Practical bayesian optimization of machine learning algorithms. In: Advances in Neural Information Processing Systems, pp. 2951–2959 (2012)

Suganthan, P.N., Hansen, N., Liang, J.J., Deb, K., Chen, Y.P., Auger, A., Tiwari, S.: Problem definitions and evaluation criteria for the CEC 2005 special session on real-parameter optimization. KanGAL Report **2005005** (2005)

Wang, Z., Jegelka, S.: Max-value entropy search for efficient Bayesian optimization. In: Proceedings of the 34th International Conference on Machine Learning, vol. 70, pp. 3627–3635. JMLR.org (2017, August)

Zhang, Y., Bahadori, M.T., Su, H., Sun, J.: FLASH: fast Bayesian optimization for data analytic pipelines.In: Proceedings of the 22nd ACM SIGKDD International Conference on Knowledge Discovery and Data Mining, pp. 2065–2074 (2016). https://doi.org/10.1145/2939672.2939829

Chapter 7
Selected Applications

It has been already pointed out in the preface that BO has been gaining increasing importance in widespread and ubiquitous application domains.

This chapter is divided into two sections. Section 7.1 gives a brief overview of several application domains where BO has enabled significant improvements over previous design and management tools. The choice of domains is driven by relatively "unconventional" and not widely known applications. The authors are aware of the arbitrariness and incompleteness of the choices given also the constraint on the length of the chapter and that even very important points might have been unwillingly skipped. The comments about software applications are intentionally short given that the theme is dealt with also in other parts of the book.

Section 7.2 is devoted to a specific analysis of the BO applications to "Smart water". The authors have been working on this subject in some European and national projects. The issues dealt with in this section are common to many problems in design and management of networked infrastructures like transportation, energy, etc. which all require failures localization, and more generally the evaluation of network resilience, demand forecasting and the integration of simulation and optimization with multi-sourced data. In the reference problem of the energy cost optimization of a real-life water distribution network, the sample efficiency of BO makes it possible to handle black box and partially defined objective functions and constraints, as introduced in Chap. 5, and to deal effectively with the issue of feasibility.

7.1 Overview of Applications

Neuroimaging: the relationship between stimuli and the neural response is analysed in (Lorenz et al. 2016) where a BO approach is suggested to balance "dynamically exploration and exploitation". The brain activity associated to different stimuli, measured by real time fMRI, is compared to the target brain state. The distribution of the brain activity is modelled as a GP which allows to formulate a Bayesian optimization model using expected improvement as acquisition function. The results quoting

F. Archetti and A. Candelieri, *Bayesian Optimization and Data Science*,
SpringerBriefs in Optimization, https://doi.org/10.1007/978-3-030-24494-1_7

(Lorenz et al. 2016) demonstrate the feasibility of a "Automatic Neuroscientist" "turning on its head how a typical fMRI experiment is carried out". The Bayesian approach can adjust the experimental conditions in real time in order to maximize similarity with a target pattern of brain activity and shows that it is possible to obtain significant reductions in the time required to scan each individual in the parameter space. Such a sample efficiency allows to tailor clinical rehabilitation therapy optimizing the stimuli according to the individual response. Other contributions to this field are Gordon et al. (2018) and Wang and Jin (2018) for the optimization of neurostimulation strategies. Other contributions to this field are Lancaster et al. (2018) and Lorenz et al. (2019).

Life sciences: BO has been used in the virtual physiological tool kit (Cooper et al. 2010) and to tune personalized models for patient vital sign monitoring (Cheng et al. 2017; Colopy et al. 2017, 2018; Cooper et al. 2010). Other applications are focused on the optimization of exoskeleton design (Gordon et al. 2018) and tissue engineering (Olofsson et al. 2018). An interesting sensor integration, driven by BO, is proposed in (Kim et al. 2017) "human-in-the-loop Bayesian optimization of wearable device parameters". The objective is to minimize the metabolic cost in unaided human subjects by tuning walking step frequencies has been accomplished in a sample efficient way, by BO resulting as well in lower inter-subject variability and energy expenditure. The following picture depicts how the BO logic is integrated into the system (Fig. 7.1).

Drug design: the discovery of novel candidate molecules with potential activity against desired targets is of central importance in the initial stages of the drug discovery. Traditionally, this was achieved through high-throughput screening (HTS) measuring in vitro simultaneously the effect of hundreds of chemical reactions set up and measured in a robotized way. HTS has largely failed to meet the expectations and this has led to the development of computational techniques and the growing importance of virtual HTS.

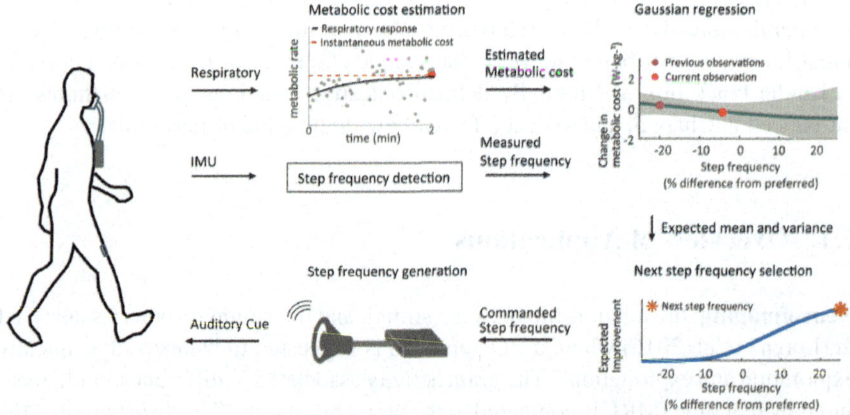

Fig. 7.1 Human in the loop Bayesian optimization, taken from (Kim et al. 2017)

Historically, this has been approached through QSAR model using cheminformatics (Kitchen et al. 2004) or machine learning (Archetti et al. 2010; Fersini et al. 2014).

Advancements in computational chemistry have enabled to calculate in silico several properties of pharmacological interest of a molecule, in particular the docking energy given a proteic receptor: however, the very size of the chemical space of candidate molecules requires an active approach in its exploration and the selection the promising ones. Virtual screening could be regarded as a computational funnel with lower fidelity surrogates being initially applied to streamline large candidate sets and more accurate-but expensive techniques being used later when the candidate pool is sufficiently small.

In this context, the sample efficiency of BO has enormous cost-saving potential in the discovery phase (Negoescu et al. 2011; Meldgaard et al. 2018; Pyzer-Knapp 2018) and clinical trials. More recently, active search methodologies have been used to improve the balance of exploitation of current data with the exploration and acquisition of new knowledge. Another approach was recently proposed by Griffiths and Hernàndez-Lobato (2017) which encodes continuous representations of molecules leveraging, in continuous latent space, gradient-based optimization to maximize some design metric.

BO is also used, (Sano et al. 2019), for pharmaceutical product development to reduce the number of experiments required to obtain the optimal formulation and process parameters.

Materials: some of the basic design problems in this field are closely related to the pharmaceutical domain: how to design, in an efficient way, molecules with significant performance metrics. A very interesting and specific reference is the volume (Packwood 2017).

The properties of new possible materials can now be calculated in silico: BO with its capacity of active learning can be applied to the analysis of huge databases of previous materials design experiments to suggest the features of new materials whose performance is evaluated by simulation in silico or through actual chemical experimentation. The final goal is to improve the design and fabrication process while reducing the number—and costs—of experimentations.

Relatively, simple examples are the prediction of compounds with low thermal conductivity (Seko et al. 2015), optimal melting temperatures (Seko et al. 2014) and elastic properties (Balachandran et al. 2016). Both (Seko et al. 2015) and (Solomou et al. 2018) focus on shape memory alloys.

In a study by (Ju et al. 2017), BO was used to design nanostructures with high thermal conductance: using the Python library COMBO from a library of 12,800 candidates structures only 438 candidate structures with excellent properties were selected for further screening.

BO has been used to compute the energy minimizing atomic structure of complex materials (Kiyohara et al. 2016).

The general problem in material design, as addressed by density functional theory (DFT), requires a general methodology for:

(a) choosing the adsorption site and the orientation for each molecule
(b) finding the energy minimizing conformation of the molecule.

The repeated execution of steps (a) and (b) could require impossibly long computational time.

Bayesian optimization is shown experimentally to provide to reduce significantly the computational time.

Active learning by Bayesian optimization in the design of new materials is also used in the paper (Lookman et al. 2019).

Structural design: BO has been used in aeronautics (Li and Pan 2019; Palar et al. 2019; Chaudhuri et al. 2019), and inspection (Liang 2018). In Lam et al. (2016), it was performed a multi-fidelity optimization using Statistical Surrogate Modelling for Non-Hierarchical Information Sources. In Basudhar et al. (2012), the geometry of a simplified aluminium wing is optimized to minimize its weigh under stability requirements. An interesting constrained BO approach, described in (Ghoreishi and Allaire 2018), has been tested on a two-dimensional aerodynamic design example. The information sources are two computational fluid dynamic simulators.

Environment: BO has been applied in Garnett et al. (2010) to optimize the predictive accuracy of a sensor network. Other applications are given in Pirot et al. (2019) for the localization of contamination sources in a discretized spatial domain. Even in cases where the geology is assumed to be perfectly known the localization process generate multi-extremal objective functions: BO is shown to be effective, even with a significant level of noise, in localizing the global minimizer which corresponds to the contaminant source location. Another relevant application is for seismic full waveform inversion (Galuzzi et al. 2018).

Energy: a paper of general interest is Dong and Chen (2019), which proposes a BO algorithm to minimize online the cost of generation, energy storage and energy purchase from the main grid. BO has also been used to improve the efficiency, i.e. optimize the power output of nano-enhanced microbial fuel cells through the optimization of the surface properties of the anodes (Azimi et al. 2012) and extended (Dolatnia et al. 2015) the optimization of the whole experimental activity at pilot scale. Other applications use BO to manage a wind and storage power plant participating in energy pool market (Crespo-Vazquez et al. 2018), the design of the controller in airborne wind energy systems (Baheri and Vermillion 2019) and the optimal estimation of the rotor effective wind speed (Moustakis et al. 2019).

Automated control: among the many applications in which BO provides data-efficient auto-tuning of control devices, we selected the calibration of the combustion engine: a fixed control of the Proportional Integral Derivative (PID) parameters can result in unstable control loops that could bring to engine to critical states. A BO approach has been proposed in Schillinger et al. (2017) to optimize "safely" the controller parameters: the loss function of the controller is modelled as a GP and the LCB acquisition function is used to sample new points. A "safe" version of Bayesian optimization has been presented in Chap. 5.

Another application of Bayesian optimization for auto-tuning is given in Neumann-brosig et al. (2018).

Robotics is another key application where BO is mostly applied in connection with reinforcement learning. (Ghosh et al. 2018). In Bansal et al. (2017), the BO is used to learn the goal-driven dynamic in robotic systems.

Recently, a safe Bayesian optimization algorithm, called SafeOpt (Chap. 5), has been developed, which guarantees that the performance of the system never falls below a critical value. In a recent paper (Marco et al. 2017), a framework is proposed where auto-tuning methods model the performance objective using standard GP models, while the specificity of the problem is represented by the covariance function. Another recent study based on GP for generating safe policy search strategies is presented in Polymenakos et al. (2019).

Software. A very significant application of BO to the tuning of software systems is reported in (Letham et al. 2019). Across the board applications of BO are reported in Golovin et al. (2017) for hyperparameter tuning. HyperTune Vizier is used across Google to optimize hyperparameters of machine learning models, both for research and production models. Vizier has reportedly made improvements to production models underlying many Google products. The HyperTune subsystem can be used also by external researchers and developers using Google Cloud Machine Learning, for instance, for automated A/B testing tuning user–interface parameters or traffic-serving parameters.

Recommender systems. The performance of the predictor of customers' ratings is highly dependent on a number of parameters, some general related to learning rate and regularization terms and some specific as the number of latent factors (Aggarwal 2016). BO has been shown to be able to tune them in a sample efficient and effective way. Another approach, referred to as *top-k recommendation problem*, formulates directly the prediction problem using the BO framework (Vanchinathan et al. 2014). Incorporating the prior information into the GP kernel allows us to make predictions about unobserved item–context pairs. Another relevant approach is Bogunovic et al. (2018) which combines a robust BO approach, namely StableOpt, and clustering to induce a degree of robustness in RS.

Mobility A number of applications to urban mobility have been implemented using tools from BO: in Zhan et al. (2018), it was performed hyperparameter optimization for traffic times-series prediction. In Turchetta et al. (2018) the calibration of agent-based transport simulations was performed through BO. An early application of the BO scheme has been the development of adaptive collective routing using GP- based dynamic congestion models (Liu et al. 2013) to minimize the collective travel time of all vehicles in the system. A key property of the approach is the ability to efficiently reason about the long-term value of exploration, which enables to balance the exploration–exploitation trade-off for entire fleet.

Finance: (Gonzalvez et al. 2019) explores the use of Gaussian processes and Bayesian optimization for two problems: to forecast the movement of the yield curve of US bonds and to improve the classical mean–variance model.

A trend following strategy is proposed depending on three variables:

1. the regularization term λ that regulates the turnover between two rebalancing dates

2. the window length that control the estimation of returns
3. the horizon time that measures the risk of assets.

The objective function considered, the Sharpe ratio, is optimized through BO.

7.2 Smart Water

7.2.1 Leakage Localization

The elements of the network are subject to failures, not uncommon given the typically old age of the infrastructure. Breakages of pipes, in particular, generate bursts and leaks which can inhibit supply service of the network (or a subnetwork) and induce substantial loss of water, with an economic loss (no-revenue water), water quality problems and unnecessary increase in water pressure and consequently energy costs.

The state of the WDN is usually monitored by a number of sensors which record flows and pressures. When a leak occurs, sensors record a variation from normal operating values. We name the vector of deviations the "signature" of the leak.

A hydraulic simulation software is used to generate a wide set of data emulating the data from sensors according to different "leakage scenarios" in the WDN, consisting in placing, in turn, a leak on each pipe and varying its severity in a given range. Our choice of simulator is EPANET 2.0, widely used for modelling WDNs and downloadable for free from the Environmental Protection Agency web site (http://www.epa.gov/nrmrl/wswrd/dw/epanet.html).

The authors have developed a machine learning-based system which, given a new leak signature, infers its location as a limited set of probably leaky pipes. Learning is performed on a dataset obtained as vectors of N components, where N is the overall number of sensors ($N = N_p + N_f$, with N_p = number of pressure sensors and N_f number of flow sensors), (also the dimension of the signature) that is the input space. According to Fig. 7.1, learning is performed in two stages: one unsupervised (i.e. kernel k-means), aimed at grouping together similar "signatures" to reduce the number of different "effects", and one supervised (i.e. SVM classification), aimed at estimating the group of signatures which the signature of a real leak belongs to (Figs. 7.2 and 7.3).

The basic idea of the analytical framework has been presented in Candelieri et al. (2015, 2017), the cluster assignment provided to each instance (i.e. signature) is used as label to train the SVM classifier. While the clustering is used to model the direct relation from leak scenario (i.e. leaky pipe and leak severity) to a group of similar signatures, the SVM inverts this relation. Thus, when a reading from sensors is acquired, the variations with respect to the faultless WDN model are computed and the resulting "signature" is given as input to the trained SVM which assigns the most probable cluster the signature belongs to. Finally, the pipes relative to the scenarios in that cluster are selected as "leaky pipes". This indication, albeit not unequivocal is much more informative than the actually used practice which yield information

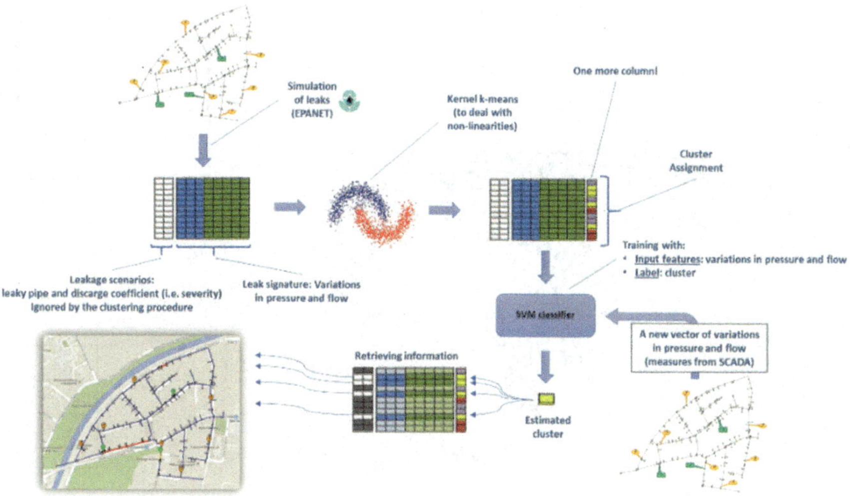

Fig. 7.2 Overall machine learning based leakage localization system. (*Source* Candelieri et al. 2017)

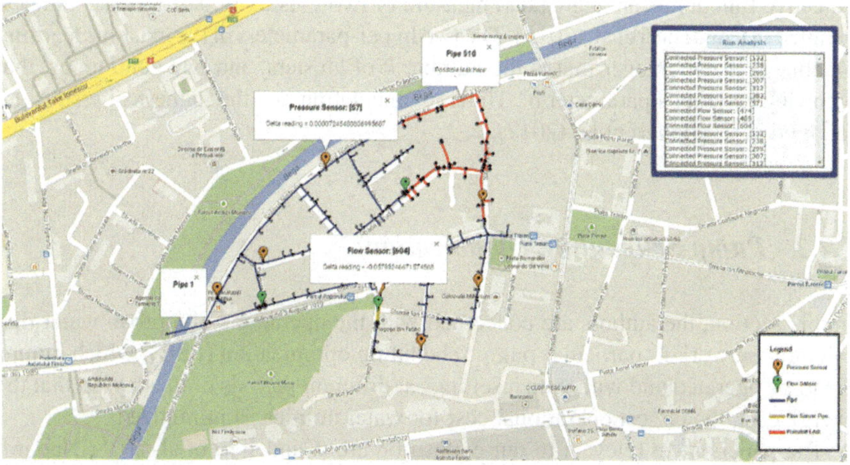

Fig. 7.3 Neptun DMA in Timisoara

at district level only. Although data can be gathered looking at historical leakage events, they would be too sparse and of poor quality. The disambiguation inside the cluster can be done "integrating" context information related to historical failure data or data pipes like age, material, etc. This leak detection solution is accessible as a web service and has been applied in a number of research and customer projects. The following Fig. 7.4 depicts a district metered area (DMA), namely "Neptun", of the urban WDN in Timisoara, Romania, was a pilot of the European Project ICeWater.

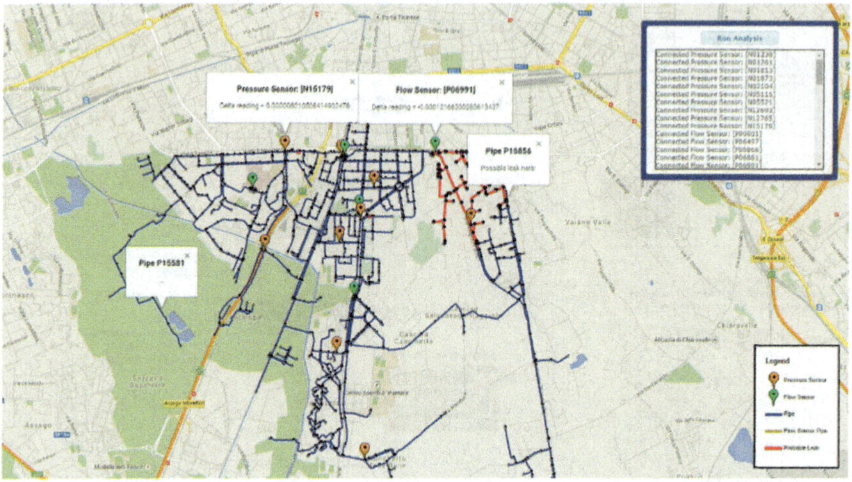

Fig. 7.4 Pilot zone "Abbiategrasso" in Milan

Neptune consists of 335 junctions (92 are consumption points) and 339 pipes. The objective function is an aggregated measure of two-clustering performance indices, detailed in Candelieri et al. (2015). The two hyper-parameters pf the machine learning pipeline are: one integer that is the number k of clusters, and one continuous, that is the RBF kernel's parameter σ. The implementation and the numerical results are reported in Candelieri et al. (2017).

7.2.2 *Pump Scheduling Optimization*

In this section, the authors are concerned with the operations of a urban water distribution network in particular pump scheduling optimization (PSO): which pumps are to be operated and with which settings at different periods of the day, so that the energy cost, the largest operational cost for water utilities, is minimized.

The papers most relevant to our exposition are (Candelieri et al. 2018; Naoum-Sawaya et al. 2015; Przystałka 2018; Tsai 2018).

We outline the complete mathematical optimization model in order to clarify the difference between the classical mathematical programming approach and BO.

The objective function is the energy cost:

$$\min C = \Delta t \sum_{t=1}^{T} c_E^t \left(\sum_{k=1}^{|\mathcal{P}|} \gamma x_{ij}^t q_{ij}^t \frac{h_{ij}^t}{\eta_{ij}} \right), \forall (i, j) \in \mathcal{P} \tag{7.1}$$

where

- $\mathcal{P}$—set of pumps, with $\mathcal{P} = \mathcal{P}_{01} \cup \mathcal{P}_{VS}$ such that $\mathcal{P}_{01} \cap \mathcal{P}_{VS} = \emptyset$, and $\mathcal{P}_{01}$ is the set of ON/OFF pumps and $\mathcal{P}_{VS}$ is the set of Variable Speed Pumps (VSP). We remark that each pump is located on a link and identified with the incident nodes.
- *C—total energy cost over T time steps*
- Δt– *time step width*
- C_E^t– *energy cost at time t*
- γ– *specific weight of water*
- q_{ij}^t– *water flow associated to a generic link (i, j), when* $(i, j) \in \mathcal{P}$, q_{ij}^t *represents the flow provided through a pump*
- h_{ij}^t– *head loss on pump* $(i, j) \in \mathcal{P}$ *at time t*
- η_{ij}– *efficiency of the pump* $(i, j) \in \mathcal{P}$

and the decision variables are:

$$x_{ij}^t \in \begin{cases} \{0, 1\} \text{ with } t = 1, \ldots, T - \text{status of pump } (i, j) \in \mathcal{P}_{01} \\ [0, 1] \text{ with } t = 1, \ldots, T - \text{speed of pump } (i, j) \in \mathcal{P}_{VS} \text{ at time } t \end{cases}$$

The optimization problem reported above is subject to the following set of constraints:

$q_{ij}^t \le \bar{q}_{ij} \, \forall (i, j) \in \mathcal{L} \backslash \mathcal{P}, t \in T$, limiting the maximum flow along each pipe $(i, j) \in \mathcal{L} \backslash \mathcal{P}$ (i.e., belonging to the set of links $\mathcal{L}$ excluding pipes $\mathcal{P}$)

$q_{ij}^t \le x_{ij}^t \bar{q}_{ij} \, \forall (i, j) \in \mathcal{P}, t \in T$, limiting the maximum flow provided by every pump $(i, j) \in \mathcal{P}$

$\sum_i q_{ij}^t - \sum_k q_{jk}^t = D_j^t \, \forall j \in J, \forall t \in T$, defining the (water) mass balance equation for each node $j \in J$, where D_j^t is the demand at node *j* at time *t*.

The following four constraints are related to the conservation of energy based on the hydraulic properties of water networks, where h_{ij}^t models the head-loss (i.e. the difference in heads) between two nodes *i* and *j* connected by a (*i*, *j*). Constraints (7.2) and (7.4) are only enforced when the corresponding q_{ij}^t are positive and are otherwise redundant.

$$q_{ij}^t \left((p_i^t + E_i) - (p_j^t + E_j) - h_{ij}^t \right) \ge 0 \, \forall (i, j) \in \mathcal{L} \backslash \mathcal{P} \forall t \in T \tag{7.2}$$

$$(p_i^t + E_i) - (p_j^t + E_j) - h_{ij}^t \le 0 \, \forall (i, j) \in \mathcal{L} \backslash \mathcal{P}, \forall t \in T \tag{7.3}$$

$$x_{ij}^t \left((p_i^t + E_i) - (p_j^t + E_j) - h_{ij}^t \right) = 0 \, \forall (i, j) \in \mathcal{P}, \forall t \in T \tag{7.4}$$

$$h_{ij}^t = \left(a_{ij} (q_{ij}^t)^2 + b_{ij} q_{ij}^t + c_{ij} \right) \forall (i, j) \in \mathcal{L}, \forall t \in T \tag{7.5}$$

In the case tanks are included in the network, we need a further constraint setting their minimum and the maximum levels.

$$\underline{P}_j^t \le p_j^t \le \overline{P}_j^t \, \forall j \in \mathcal{T}, \text{ that is the set of tanks, and } t \in T$$

The following constraint is the tank mass balance equation where A_j is the surface area of tank $j \in \mathcal{T}$.

$$p_j^t = p_j^{t-1} + \frac{\sum_i q_{ij}^t - \sum_k q_{jk}^t}{\Delta t\, A_j} \forall j \in \mathcal{T}, \forall t \in T$$

Finally, the following constraint denotes the binary and non-negativity conditions.

$$q_{ij}^t \geq 0 \forall (i, j) \in \mathcal{P}$$

In this case, the decision variables of the optimization problem are: the pump schedule x_{ij}^t, the head-loss h_{ij}^t and the flow for every pipe and pump q_{ij}^t.

The goal of PSO is to identify the pump schedule, consisting in the status of each pump, associated with the lowest energy cost while satisfying hydraulic feasibility (i.e. water demand satisfaction, pressure level within a given operational range, etc.) The status of a pump is its activation—in the case of an ON/OFF pump—or its speed—in the case of a VSP, leading to discrete or continuous decision variables, respectively. Nonlinearities, binary and continuous decision variables make PSO computationally challenging, even for small WDNs.

For a survey of the optimization methods employed (Mala-Jetmarova et al. 2017). Although the PSO objective function (7.1) has an apparently closed expression, its exact resolution through mathematical programming requires to approximate the flow equations (linearization or convexification) (Pecci et al. 2017) which regulates the hydraulic behaviour of a pressurized WDN, according to constraints (7.2–7.5).

The time horizon usually adopted in PSO is one day with hourly time steps, leading to $T = 24$ time steps overall. This choice is basically motivated by the energy tariffs which are typically characterized by a different energy price depending on the hour of the day.

The optimal solution found by the mathematical programming algorithm is then tested for feasibility through a hydraulic simulator. In case of unfeasible solution, feasibility should be restored heuristically via local search or by improving the accuracy of mathematical model.

As the size of the network increases, the number of decision variables (i.e. the pump schedule x_{ij}^t, the head-loss h_{ij}^t and the flow for every pipe and pump q_{ij}^t) grows leading to increase in the number of function evaluations, computational resources (simulation time increases with the number of hydraulic components of the WDN) and memory requirements. Since the 1990s metaheuristics, such as genetic algorithms and simulated annealing, have been increasingly applied given their versatility and a problem-independent global approach that can be applied to complex problems: anyway, in large search spaces, their slow convergence results in far too long computations. Moreover, evolutionary algorithms do not cope well with constraints. In all the approaches, it is important to reduce the computational load of hydraulic simulation: one way to do this is working on the mathematical model of the hydraulic components through linearization/convexification of objective or constraints, the other uses the sample efficiency of BO to limit the number of simulation runs. We remark that we use EPANET for every function evaluation (Candelieri et al. 2018) of the full model of the WDN, while mathematical programming uses EPANET only ex-post to evaluate the feasibility of the optimal solution of the "simplified" model. The proposed BO approaches are compared (Candelieri et al. 2018) to two commercial products: an "optimization-as-a-service" cloud-based platform (SigOpt) and the combination of two commercial products: ALAMO and BARON for the generation of an accurate surrogate model and the successive optimization process. A further approach, namely "Feasibility Set Approximation Probabilistic Branch and Bound", has been recently proposed to address the problem as a black-box feasibility determination. The new stochastic partitioning algorithm is based on an extension of probabilistic branch and bound (PB&B) to properly approximate the feasible set (Tsai et al. 2018). The algorithm iteratively maintains, prunes and branches subregions to approximate the unknown target feasibility set with a probabilistic guarantee of the accuracy. When the algorithm terminates, it provides subregions with a relatively high concentration of high-quality solutions and returns an estimation of the global optima.

In Candelieri et al. (2018), the BO approach and the combination of two commercial products: ALAMO and BARON have been compared under the same experimental setting, considering a real-life large-scale WDN in Milan, Italy. During the European Project ICeWater, a pressure management zone (PMZ), namely "Abbiategrasso", was identified in the South of the city and isolated from the rest of the network. The zone is served by a pump station with two pumps.

The objective function is partially defined depending on the computation of the constraints, which are:

- Black box: because their evaluation requires to run EPANET
- Dependent. To give an example, let us consider three simple constraints violations: (1) water demand in a node is not satisfied, (2) pressure is negative in at least a junction and (3) one of the tanks is empty. For these reasons, approaches like ICEI, introduced Chap. 5 cannot be applied.

To solve PSO, we have applied two schemes, the first (Candelieri et al. 2018) assigns a fixed "penalty" as value of the objective function for evaluations outside the feasible region. The second approach (i.e. SVM-CBO), presented in Chap. 5, does not require constraints independence and consists in two consecutive phases: the estimation of the feasible region and the BO constrained on such an estimate.

The number of overall function evaluations (i.e. EPANET simulation runs) has been fixed at 1200, organized as follows, for the SVM-CBO:

- 400 for the initial design, via LHS
- 400 for phase 1 (feasibility determination)
- 400 for phase 2 (Constrained Bayesian optimization on the estimate of the feasible region).

As far as the BO with penalty is concerned, the overall budget is dived in:

- 400 for initial design, via LHS
- 800 for the BO process.

To make significant, the comparison between the two schemes, penalty and SVM-CBO, they share the same initial design. They also share all the other choices, more precisely the probabilistic surrogate model (GP) and the acquisition function (LCB), so that the difference in the result can be only motivated by the usage, or not, of the estimate of the feasible region. Since the value $f(x^*)$ is not known for this case study, the Gap metrics cannot be used as metric, differently from the test functions reported in Chap. 5. The minimum value of the objective function observed at the end of the approaches, that is the energy cost associated to the best seen pump schedule, has been directly considered as metric.

As already reported by the authors in their previous work (Candelieri et al. 2018), assigning a penalty to function evaluations outside the feasible region has the aim to move towards feasible points/regions. Although this is usual the usual solution adopted, it is quite naïve in this problem especially when the penalty cannot be made dependent on the entity of violation. This leads to almost "flat" objective functions, especially when the unknown feasible region results extremely narrow with respect to the overall search space. As already reported, the real-life PSO case study seems to be characterized by this kind of situation.

BO with penalty was not able to provide any improvement with respect to the energy cost identified on the initial design that is 172.89€. On the contrary, SVM-CBO was able to further reduce the energy to 168.60€, by identifying a new and more efficient pump schedule. A significantly relevant result is that the improvement has been registered exactly at the starting of phase 2, just shown for the test functions in Chap. 5. This result proves that solving feasibility determination and constraining the following sequential optimization process to an accurate estimate of the feasible region can significantly improve the effectiveness and efficiency of the BO process.

A comment is due about the optimal value which is significantly lower than the average pumping cost because the pilot zone was equipped with VSP's and monitored and maintained specifically for the european project.

References

Aggarwal, C.C.: Recommender systems, pp. 1–28. Springer International Publishing, Cham (2016)

Archetti, F., Giordani, I., Vanneschi, L.: Genetic programming for QSAR investigation of docking energy. Appl. Soft Comput. **10**(1), 170–182 (2010)

Ashmaig, O., Connolly, M., Gross, R.E., Mahmoudi, B.: Bayesian optimization of asynchronous distributed microelectrode theta stimulation and spatial memory. In: 2018 40th Annual International Conference of the IEEE Engineering in Medicine and Biology Society (EMBC), pp. 2683–2686 (2018)

Azimi, J., Jalali, A., Fern, X.: Hybrid batch Bayesian optimization 1–12 (2012). doi:10.1.1.467.8687. arXiv preprint arXiv:1202.5597

Baheri, A., Vermillion, C.: Combined plant and controller design using batch Bayesian optimization: a case study in airborne wind energy systems. J. Dyn. Syst. Meas. Control (2019)

Balachandran, P.V., Xue, D., Theiler, J., Hogden, J., Lookman, T.: Adaptive strategies for materials design using uncertainties. Scientific Reports. **6** (2016). https://doi.org/10.1038/srep19660

Bansal, S., Calandra, R., Xiao, T., Levine, S., Tomiin, C.J.: Goal-driven dynamics learning via Bayesian optimization. In: 017 IEEE 56th Annual Conference on Decision and Control (CDC), pp. 5168–5173. IEEE (2017)

Bogunovic, I., Scarlett, J., Jegelka, S., Cevher, V.: Adversarially Robust Optimization with Gaussian Processes. In: Advances in Neural Information Processing Systems, pp. 5760–5770 (2018)

Candelieri, A., Giordani, I., Archetti, F.: Automatic configuration of Kernel-based clustering: an optimization approach. In: International Conference on Learning and Intelligent Optimization, pp. 34–49. Springer, Cham (2017, June)

Candelieri, A., Soldi, D., Archetti, F.: Cost-effective sensors placement and leak localization—The Neptune pilot of the ICeWater project. J. Water Supply Res. Technol. AQUA. 567–582 (2015)

Candelieri, A., Perego, R., Archetti, F.: Bayesian optimization of pump operations in water distribution systems. J. Global Optim. **71**, 213–235 (2018). https://doi.org/10.1007/s10898-018-0641-2

Chaudhuri, A., Marques, A.N., Lam, R., Willcox, K.: Reusing Information for Multifidelity Active Learning in Reliability-Based Design Optimization. AIAA Scitech 2019 Forum. 1–12 (2019). https://doi.org/10.2514/6.2019-1222

Cheng, L.-F., Darnell, G., Dumitrascu, B., Chivers, C., Draugelis, M.E., Li, K., Engelhardt, B.E.: Sparse multi-output gaussian processes for medical time series prediction (2017). arXiv preprint arXiv:1703.09112

Colopy, G.W., Pimentel, M.A.F.F., Roberts, S.J., Clifton, D.A.: Bayesian optimisation of Gaussian processes for identifying the deteriorating patient. In: 2017 IEEE EMBS International Conference on Biomedical and Health Informatics, BHI 2017. pp. 85–88 (2017)

Colopy, G.W., Roberts, S.J., Clifton, D.A.: Bayesian optimization of personalized models for patient vital-sign monitoring. IEEE J. Biomed. Health Inform. **22**, 301–310 (2018). https://doi.org/10.1109/JBHI.2017.2751509

Cooper, J., Cervenansky, F., De Fabritiis, G., Fenner, J., Friboulet, D., Giorgino, T., Manos, S., Martelli, Y., Villá-Freixa, J., Zasada, S., Lloyd, S., Mccormack, K., Coveney, P.V.: The virtual physiological human toolkit. Philos. Trans. Royal Soc. A Mathe. Phys Eng. Sci. **368**, 3925–3936 (2010). https://doi.org/10.1098/rsta.2010.0144

Crespo-Vazquez, J.L., Carrillo, C., Diaz-Dorado, E., Martinez-Lorenzo, J.A., Noor-E-Alam, M.: A machine learning based stochastic optimization framework for a wind and storage power plant participating in energy pool market. Appl. Energy **232**, 341–357 (2018). https://doi.org/10.1016/j.apenergy.2018.09.195

Dolatnia, N., Fern, A., Fern, X.: Toward embedding Bayesian optimization in the lab: reasoning about resource and actions. In: AAAI Fall Symposium—Technical Report (2015)

Dong, G., Chen, Z.: Data-driven energy management in a home microgrid based on Bayesian optimal algorithm. IEEE Trans. Industr. Inf. **15**(2), 869–877 (2019)

Fersini, E., Messina, E., Archetti, F.: A p-median approach for predicting drug response in tumour cells. BMC Bioinform. **15** (2014). https://doi.org/10.1186/s12859-014-0353-7

Galuzzi, B.G., Perego, R., Candelieri, A., Archetti, F.: Bayesian optimization for full waveform inversion. In: Daniele P., S.L. (eds.) New Trends in Emerging Complex Real Life Problems, pp. 257–264. Springer International Publishing, Taormina (2018)

Garnett, R., Osborne, M.A., Roberts, S.J.: Bayesian optimization for sensor set selection. In: Proceedings of the 9th ACM/IEEE International Conference on Information Processing in Sensor Networks—IPSN'10, pp. 209–219. Stockholm (2010)

Ghoreishi, S.F., Allaire, D.: Multi-information source constrained Bayesian optimization. Struct. Multidisciplinary Optim. 1–15 (2018). https://doi.org/10.1007/s00158-018-2115-z

Ghosh, S., Berkenkamp, F., Ranade, G., Qadeer, S., Kapoor, A.: Verifying Controllers Against Adversarial Examples with Bayesian Optimization. In: IEEE International Conference on Robotics and Automation (ICRA). pp. 7306–7313, Brisbane (2018)

Golovin, D., Solnik, B., Moitra, S., Kochanski, G., Karro, J., Sculley, D.: Google Vizier: A Service for Black-Box Optimization. In: Proceedings of the 23rd ACM SIGKDD International Conference on Knowledge Discovery and Data Mining. pp. 1487–1495. ACM (2017)

Gonzalvez, J., Lezmi, E., Roncalli, T., Xu, J.: Financial applications of Gaussian processes and Bayesian optimization (2019). arXiv preprint arXiv:1903.04841

Gordon, D.F.N.N., Matsubara, T., Noda, T., Teramae, T., Morimoto, J., Vijayakumar, S.: Bayesian Optimisation of Exoskeleton Design Parameters. In: Proceedings of the IEEE RAS and EMBS International Conference on Biomedical Robotics and Biomechatronics, pp. 653–658 (2018)

Griffiths, R.-R., Hernández-Lobato, J.M.: Constrained Bayesian optimization for automatic chemical design (2017). arXiv preprint arXiv:1709.05501

Ju, S., Shiga, T., Feng, L., Hou, Z., Tsuda, K., Shiomi, J.: Designing nanostructures for phonon transport via Bayesian optimization. Phys. Rev. X **7**(2), 021024 (2017)

Kim, M., Ding, Y., Malcolm, P., Speeckaert, J., Siviy, C.J., Walsh, C.J., Kuindersma, S.: Human-in-the-loop Bayesian optimization of wearable device parameters. PloS one. **12**(9), e0184054 (2017)

Kitchen, D.B., Decornez, H., Furr, J.R., Bajorath, J.: Docking and scoring in virtual screening for drug discovery: methods and applications. Nat. Rev. Drug Discov. **3**(11), 935 (2004)

Kiyohara, S., Oda, H., Tsuda, K., Mizoguchi, T.: Acceleration of stable interface structure searching using a kriging approach. Japanese J. Appl. Phys. **55** (2016). https://doi.org/10.7567/jjap.55.045502

Lam, R., Willcox, K., Wolpert, D.H.: Bayesian optimization with a finite budget: an ap-proximate dynamic programming approach. Adv. Neural. Inf. Process. Syst. **30**, 883–891 (2016). https://doi.org/10.1186/1471-2407-4-76

Lancaster, J., Lorenz, R., Leech, R., Cole, J.H.: Bayesian optimization for neuroimaging pre-processing in brain age classification and prediction. Front. Aging Neuro. **10**, 28 (2018)

Letham, B., Karrer, B., Ottoni, G., Bakshy, E.: Constrained Bayesian optimization with noisy experiments. Bayesian Anal. **14**(2), 495–519 (2019)

Liang, X.: Image-based post-disaster inspection of reinforced concrete bridge systems using deep learning with Bayesian optimization. Comput.-Aided Civil Infrastruct. Eng (2018). https://doi.org/10.1111/mice.12425

Li, C., Pan, Q.: Adaptive optimization methodology based on Kriging modeling and a trust region method. Chinese J. Aeronaut. (2019). https://doi.org/10.1016/j.cja.2018.11.012

Liu, S., Yue, Y., Krishnan, R.: Adaptive collective routing using gaussian process dynamic congestion models. In: Proceedings of the 19th ACM SIGKDD International Conference on Knowledge Discovery and Data Mining, pp. 704–712. ACM (2013, August)

Lookman, T., Balachandran, P.V., Xue, D., Yuan, R.: Active learning in materials science with emphasis on adaptive sampling using uncertainties for targeted design. Computat. Mater. **5**, 21 (2019). https://doi.org/10.1038/s41524-019-0153-8

Lorenz, R., Monti, R.P., Violante, I.R., Anagnostopoulos, C., Faisal, A.A., Montana, G., Leech, R.: The automatic neuroscientist: a framework for optimizing experimental design with closed-loop real-time fMRI. NeuroImage **129**, 320–334 (2016). https://doi.org/10.1016/j.neuroimage.2016.01.032

Lorenz, R., Simmons, L.E., Monti, R.P., Arthur, J.L., Limal, S., Laakso, I., Leech, R., Violante, I.R.: Efficiently searching through large tACS parameter spaces using closed-loop Bayesian optimization. Brain Stimul. (2019)

Mala-Jetmarova, H., Sultanova, N., Savic, D.: Lost in optimisation of water distribution systems? A literature review of system operation. Environ. Model Softw. **93**, 209–254 (2017)

Marco, A., Berkenkamp, F., Hennig, P., Schoellig, A. P., Krause, A., Schaal, S., Trimpe, S.: Virtual versus real: trading off simulations and physical experiments in reinforcement learning with bayesian optimization. In: 2017 IEEE International Conference on Robotics and Automation (ICRA), pp. 1557–1563. IEEE (2017, May)

Meldgaard, S.A., Kolsbjerg, E.L., Hammer, B.: Machine learning enhanced global optimization by clustering local environments to enable bundled atomic energies. J. Chem. Phys. **149** (2018). https://doi.org/10.1063/1.5048290

Moustakis, N., Mulders, S. P., Kober, J., van Wingerden, J.W.A.: Practical Bayesian optimization approach for the optimal estimation of the rotor effective wind speed (2019)

Naoum-Sawaya, J., Ghaddar, B., Arandia, E., Eck, B.: Simulation-optimization approaches for water pump scheduling and pipe replacement problems. Eur. J. Oper. Res. **246**, 293–306 (2015). https://doi.org/10.1016/j.ejor.2015.04.028

Negoescu, D.M., Frazier, P.I., Powell, W.B.: The knowledge-gradient algorithm for sequencing experiments in drug discovery. INFORMS J. Comput. **23**, 346–363 (2011). https://doi.org/10.1287/ijoc.1100.0417

Neumann-brosig, M., Marco, A., Schwarzmann, D., Trimpe, S.: Data-efficient Auto-tuning with Bayesian optimization: an industrial control study (2018). arXiv preprint arXiv:1812.06325

Olofsson, S., Mehrian, M., Calandra, R., Geris, L., Deisenroth, M., Misener, R.: Bayesian multi-objective optimisation with mixed analytical and black-box functions: application to tissue engineering, (2018)

Packwood, D.: Bayesian Optimization for Materials Science. Springer (2017)

Palar, P.S., Dwianto, Y.B., Regis, R.G., Oyama, A., Zuhal, L.R.: Benchmarking Constrained Surrogate-based Optimization on Low Speed Airfoil Design Problems (2019)

Pecci, F., Abraham, E., Stoianov, I.: Quadratic head loss approximations for optimisation problems in water supply networks, J. Hydroinformatics. **19**, 493–506 (2017). ISSN: 1464-7141

Pirot, G., Krityakierne, T., Ginsbourger, D., Renard, P.: Contaminant source localization via Bayesian global optimization. Hydrol. Earth Syst. Sci. **23**(1) (2019)

Polymenakos, K., Abate, A., Roberts, S.: Safe policy search using Gaussian process models. In: Proceedings of the 18th International Conference on Autonomous Agents and MultiAgent Systems, pp. 1565–1573. International Foundation for Autonomous Agents and Multiagent Systems (2019, May)

Przystałka, P.: Performance optimization of a leak detection scheme for water distribution networks. IFAC-PapersOnLine **51**(24), 914–921 (2018)

Pyzer-Knapp, E.O.: Bayesian optimization for accelerated drug discovery. IBM J. Res. Develop. 1–1 (2018). https://doi.org/10.1147/jrd.2018.2881731

Sano, S., Kadowaki, T., Tsuda, K., Kimura, S.: Application of bayesian optimization for pharmaceutical product development. J. Pharmaceutical Innov. 1–11 (2019)

Schillinger, M., Hartmann, B., Skalecki, P., Meister, M., Nguyen-Tuong, D., Nelles, O.: Safe active learning and safe Bayesian optimization for tuning a PI-controller. IFAC-PapersOnLine **50**, 5967–5972 (2017). https://doi.org/10.1016/j.ifacol.2017.08.1258

Seko, A., Maekawa, T., Tsuda, K., Tanaka, I.: Machine learning with systematic density-functional theory calculations: Application to melting temperatures of single- and binary-component solids. Phys. Rev. B Condens. Matter Mater. Phys. **89** (2014). https://doi.org/10.1103/physrevb.89.054303

Seko, A., Togo, A., Hayashi, H., Tsuda, K., Chaput, L., Tanaka, I.: Prediction of low-thermal-conductivity compounds with first-principles anharmonic lattice-dynamics calculations and bayesian optimization. Phys. Rev. Lett. **115** (2015). https://doi.org/10.1103/physrevlett.115.205901

Solomou, A., Zhao, G., Boluki, S., Joy, J.K., Qian, X., Karaman, I., Arroyave, R., Lagoudas, D.C.: Multi-objective bayesian materials discovery: Application on the discovery of precipitation strengthened NiTi shape memory alloys through micromechanical modeling. Mater. & Des. **160**, 810–827 (2018)

Tsai, Y.-A., Pedrielli, G., Mathesen, Logan Zabinsky, Z.B., Huang, H., Candelieri, Antonio Perego, R.: Stochastic optimization for feasibility determination: an application to water pump operation in water distribution network. In: Proceedings of the 2018 Winter Simulation Conference, pp. 1945–1956. IEEE, Gothenburg, Sweden (2018)

Turchetta, M., Makarova, A., Beyeler, S.: Calibration of agent based transport simulations with multi-fidelity Bayesian optimization. In: 18th Swiss Transport Research Conference (STRC 2018) (2018)

Vanchinathan, H.P., Nikolic, I., De Bona, F., Krause, A.: Explore-exploit in top-N recommender systems via Gaussian processes. In: Proceedings of the 8th ACM Conference on Recommender systems—RecSys'14, pp. 225–232 (2014)

Wang, H., Jin, Y.: A random forest-assisted evolutionary algorithm for data-driven constrained multiobjective combinatorial optimization of trauma systems. IEEE Trans. Cybern (2018). https://doi.org/10.1109/tcyb.2018.2869674

Zhan, H., Gomes, G., Li, X.S., Madduri, K., Wu, K.: Efficient online hyperparameter optimization for Kernel ridge regression with applications to traffic time series prediction. **89**, 1–19 (2018). arXiv preprint arXiv:1811.00620

Made in the USA
Las Vegas, NV
28 December 2024

15483524R00079